Materials Science and Technologies

nova
science publishers
www.novapublishers.com

Materials Science and Technologies

A Guide to Laser Welding
António Manuel de Bastos Pereira, PhD (Editor)
Eva Soledade Vaz Marques, MSc (Editor)
Francisco José Gomes da Silva (Editor)
2024. ISBN: 979-8-88697-613-7 (Softcover)
2023. ISBN: 979-8-89113-401-0 (eBook)

Biocomposites: Advances in Research and Applications
James R. Bush (Editor)
2023. ISBN: 979-8-89113-267-2 (Softcover)
2023. ISBN: 979-8-89113-313-6 (eBook)

Ferrite Materials and Technologies
Ravi Panwar, PhD (Editor)
Dharmendra Singh (Editor)
2023. ISBN: 979-8-89113-086-9 (Hardcover)

Perovskite Solar Cells: From Materials Science to Device Engineering
Thembinkosi Donald Malevu, PhD (Editor)
2023. ISBN: 979-8-89113-130-9 (Softcover)
2023. ISBN: 979-8-89113-188-0 (eBook)

Physics and Mechanics of New Materials and Their Applications, 2021 – 2022
Ivan A. Parinov, DrSc (Editor)
Shun-Hsyung Chang, PhD (Editor)
Arkady N. Soloviev, DrSc (Editor)
2023. ISBN: 979-8-88697-542-0 (eBook)

More information about this series can be found at
https://novapublishers.com/product-category/series/materials-science-and-technologies/

Pranjal Saikia
Jayanta Kumar Sarmah
and Parikshit Gogoi
Editors

Bimetals

Formation, Properties and Applications

nova
science publishers
www.novapublishers.com

Library of Congress Cataloging-in-Publication Data

ISBN: 979-8-89113-496-6

Published by Nova Science Publishers, Inc. † New York

Contents

Preface

Bimetals have been attracting significant attention in materials science due to their unique properties and wide-ranging applications. Bimetals are two different metals that are bonded together to form a composite material with superior properties than either of the individual metals. The combination of the different metals results in a material with improved mechanical, thermal, electrical, and magnetic properties, making bimetals highly sought after in various industries.

The idea of bimetals dates back to ancient times, where people used them to make tools and weapons. However, the development of bimetallic technology and its applications have advanced significantly in recent years, thanks to the remarkable progress in materials science and engineering. This book, "Bimetals: Formation, Properties, and Applications," provides an in-depth overview of the current state of the art of bimetallic materials. It covers various aspects of bimetal formation, properties, and applications, including the underlying principles, the different synthesis methods, the characterization techniques, and the wide range of applications of bimetals. The book is divided into 8 chapters, each providing an in-depth look at different aspects of bimetals.

This book is an essential reference for materials scientists, researchers, engineers, and students interested in bimetals and their applications. We hope that this book will provide an in-depth understanding of bimetals and their potential to revolutionize various R&D processes and industries.

Prof. Pranjal Saikia
Dr. Jayanta K. Sarmah
Dr. Parikshit Gogoi

Chapter 1

Shape-Controlled Bimetallic Nanostructures for Electrochemical Oxidation of Liquid Fuels

Naresh Nalajala*

Department of Materials Science and Catalysis,
Poornaprajna Institute of Scientific Research (PPISR), Devanahalli, Bangalore, India

Abstract

The growing depletion and consumption of fossil fuels highlight the urgent need for alternative fuels that are not only renewable in nature but also contribute to a low carbon footprint. In this perspective, hydrogen is recognised as a possible fuel that may be created from water in a variety of ways while simultaneously providing society with pollution-free emissions during combustion or conversion. However, not only is its manufacturing complex, but the flammability and light weight of H_2 make storage and transportation challenging, limiting the commercial success of possible H_2 uses. The scientific community has discovered various liquid fuels (LFs) that can be transformed into energy (electricity) using direct liquid fuel cells (DLFCs) while looking for alternative energy fuels. The favorable characteristics of LFs, including their renewable nature, high solubility, high energy density, low environmental impact, ease of transportation and storage, and low cost, have led to their consideration as future alternate fuels. Methanol, ethanol, formic acid, glycerol, ethylene glycol, $NaBH_4$, ammonium borane, and other fuels are used and can be converted to power by oxidation at the anode side of fuel cells. Among the efforts devoted to the development of potential catalysts, shape-engineering of active electrocatalysts is a promising approach because the surface is exposed

* Corresponding Author's Email: naresh@ppisr.res.in.

In: Bimetals
Editors: Pranjal Saikia, Jayanta Kumar Sarmah and Parikshit GogoiName
ISBN: 979-8-89113-496-6

with highly active sites towards efficient reaction kinetics and thus effective utilization of active metal (for example, Pt). Furthermore, shape engineering of bimetallic-based electrocatalysts (Pt, Pd, Ag, Ni, Co, Cu, Fe, Pb, etc.) was discovered to improve performance due to both electrical and geometric aspects. With this motivation, the purpose of this chapter is to provide information on the preparation of shape-controlled bimetallic alloy catalysts and their structure-property-activity linkages for electrochemical oxidation of LFs. This chapter is dedicated to learning the design and performance aspects of shape-controlled bimetallic catalysts based on various fuels such as methanol, ethanol, formic acid, and ethylene glycol for various DLFCs applications; however, other LFs fuels ($NaBH_4$, ammonium borane, and so on) also serve the purpose but are not addressed due to a lack of relevant literature.

Keywords: shape-control, nanostructures, bimetallics, fuel cells, liquid fuels

1. Introduction

A fuel cell is an electrochemical energy conversion device that can convert chemical energy to electricity with practically negligible emission of pollutants. To reduce dependence on finite fossil fuel resources such as oil and gas commodities, the development of fuel cell technology is crucial [1-3]. Combustion of these fuels produces greenhouse gas emissions (CO_2, CO, SO_X, NO_X, etc.). Proton exchange membrane fuel cells (PEMFCs) are promising because they can convert hydrogen, a high-energy-density fuel, into power and water; a renewable byproduct. However, the commercial success of this technique is hampered by the manufacture, transportation, and storage of H_2 owing to its easily combustible and low weight [4, 5]. Several liquid fuels (LFs) are being eyed as potential replacements for H_2 because of their renewable nature, high solubility, high energy density, low environmental contamination, simple transportation and storage requirements, and low cost. Due to their low toxicity, high boiling point, high energy density, and less pronounced crossover [6-9], a wide variety of carbon based organic (ethyl glycol and glycerol 1, 2, propanol, dimethylether) and carbon free/inorganic (hydrazine, ammonia borane, and sodium borohydride) fuels are used as LFs. The fuel (LFs) is provided at the anode, where it undergoes electrochemical oxidation to generate electrons and oxidized products (such as formaldehyde, formic acid, acetaldehyde, CO, etc.) with carbon dioxide (CO_2) as the predicted final result. Figure 1 shows the overall design of a DLFC, in which protons, electrons (passed through from the load), and oxygen all come together at the cathode to form water (via the oxygen reduction reaction (ORR)).

The main objective of this chapter is to enable readers to comprehend LFs oxidation under electrochemical conditions by highlighting important features of catalyst design for the realization of efficient DLFCs technologies. Electrochemical oxidation of LFs at an anode, with the aid of an active electrocatalyst, generates electrons and electricity in a manner analogous to that of H_2. The electrochemical oxidation of any liquid fuel undergoes several intermediate steps before producing highly stable molecules (for instance CO_2) which cannot further oxidize under working potentials (<0.9 V) of fuel cells. Due to their high activity and stability under fuel cell operation circumstances [8, 9], Pt and Pt-based materials are the only prospective electrocatalysts that could make fuel cell technology more commercially viable and sustainable. However, abundance, high cost, and poor resistance to CO poisoning drive the research for alternate catalyst materials for the electrochemical oxidation of liquid fuels [3]. A lot of effort is being invested in exploring alternatives to platinum, such as carbon-based transition metals, as catalysts, but so far none of these have reached the levels of activity and stability desired by the Department of Energy (DOE) [10-13]. However, some methods are reported to have used the active metal Pt with optimum electronic, geometric, and surface properties, leading to significantly improved activities and stability. These methods include (i) alloying Pt with transition metal elements (Fe, Ni, Co, Cu, Zn), (ii) core-shell design, in which the active metal is only confined to the surface, exposing Pt to the greatest extent possible, and (iii) shape-controlled Pt nanoparticles with exposed active sites [14-17]. Among the approaches reported for better electrochemical oxidation of LFs, the chapter is aimed at exploring the efforts devoted in shape-controlled bimetallic nanocrystals. It is well known that single crystal surfaces or well-defined surfaces are very much surface sensitive to electrochemical reaction kinetics; for instance, methanol oxidation is very much efficient on Pt (100) surface compared to Pt (111), while ethanol oxidation is superior over the Pt (111) than other surfaces Pt (755) Pt (322) [18, 19]. However, the well-defined surfaces or single crystal surfaces are limited to lab-scale studies and these metal surfaces cannot be used for practical applications because of poor surface area and design limitations with actual device applications. Since shape-controlled metal nanocrystals have many exposed active sites and a high surface-to-volume ratio [20, 21], they are useful in translating lab-scale discoveries into real-world applications. Effective electrocatalysts for a wide range of catalytic, electrocatalytic, and photocatalytic activities, including the electrochemical oxidation of LFs, have been described for numerous shape-controlled metal nanocrystals (Pt, Pd) and their alloy-based materials [22, 23]. Shape-controlled bimetallic nanoarchitectures, such as wires, cubes, sheets, octahedra, concave nanocubes, icosahedra, nanoframes, hexapods, and high index facets of different compositions PtPd, PtCu, PtFe, PtCo, PtNi, PtRu, are viewed as potential catalysts towards achieving enhanced activities and stabilities. Electrochemical oxidation of LFs is discussed, along with the use of several techniques documented for synthesizing shape-controlled bimetallic nanoparticles. Based on a comprehensive literature search, we understand that various bimetallic nanocrystal forms can function as effective electrocatalysts for the electrochemical oxidation of methanol, ethanol, formic acid, and ethylene glycol. Nevertheless, there are potential fuel candidates for DLFCs that we are unable to elaborate due to a lack of pertinent research findings. These include 1, 2-propanol, dimethylether, hydrazine, ammonia borohydride, and sodium borohydride. This chapter is divided into the following sections, which discuss the preparation methods of bimetallic nanostructures of various shapes, their applications to electrochemical oxidation of LFs (methanol, ethanol, formic acid, and ethylene glycol), and the results obtained in terms of activity (mass activity [MA] and specific activity [SA]), stability, and resistance to CO poisoning.

2. Preparation Methods

Shape-controlled metal nanocrystals are prepared with desired size and composition by employing several strategies such as seeded growth, hydrothermal, solvothermal, underpotential deposition, sacrificial templating, galvanic replacement, chemical etching, among others. For the synthesis of any shape-controlled counterparts, the use of reducing agent, stabilizer, capping agent and supporting solvents is inevitable. It is evident that by fine tuning of various experimental conditions such as temperature, time, concentration of components (for eg. metal precursor, reducing agent and capping agent), change in precursor and speed of its addition, it is possible to produce various types of shapes of bimetallic nanocrystals [24-26]. After the synthesis at room temperature, the collected nanocrystals will be subjected to further cleaning of adsorbed impurities from the surface; otherwise, the actual activities and stabilities of prepared nanocrystals cannot be achieved due to occupation of active sites with residual impurities [27, 28]. This section addresses some of the relevant literature based on the synthesis of shape-controlled bimetallic nanostructures. Additionally, for a more in-depth understanding, readers are provided with insightful references [24-26]. It is a challenging task to synthesize nanoparticles of bi- and multimetallic in nature with controlled size and shape due to the fact that different reduction potentials cause unique nucleation and growth kinetics of individual counterparts. However, have discovered discovered that through a single step synthesis method, it is possible to prepare not only bimetallic but also multimetallic nanoparticles that are bound with high index or defect structures. Zhang P et al., reported a single step/one pot solvothermal synthesis for the preparation of Pt-Ni-Cu nanoparticles bounded by high index facets. It was observed that, the shape evolution of rough octahedreal nanocrystals forms crossed, dendritic and concave nanocubic shapes of Pt-Ni-Cu particles [29]. One pot synthesis of 3D based nanoframes of PtCu of different compositions and morphologies can be prepared by following solvothermal method using appropriate precursor materials (Pt $(acac)_2$ and Cu $(acac)_2$) and other reagents [30]. With proper tuning of experimental conditions, 3D Pt_3Co nanowires assemblies (NWAs) were prepared [31]. Additionally, there were reports on the preparation of icosahedra with twin defects and defect-free octahedral Pt_3Cu nanocrystals. This was achieved by maintaining the same experimental conditions, with the only alteration being the substitution of $CuCl_2$ as a precursor for icosahedral Pt_3Cu and $Cu(acac)_2$ for octahedral Pt_3Cu. [32]. In another study, the critical role of $CuCl_2$ was established for the synthesis of highly open rhombic dodecahedral PtCu nanoframes; the production of nanoframes might be possible due to coordination interaction of chloride ions to copper ions, that enables the facile galvanic replacement between Cu seed nanocrystals and Pt species, and thereby allowing the efficient diffusion of Cu atoms [33]. For the synthesis of screw thread type PtCu nanowires, precursors ($Pt(acac)_2$, $CuCl_2.2H_2O$), reducing agent (glucose), capping agent (cetyltrimethylammonium chloride (CTAC)), solvent (oleylamine) being used [34]. It was also reported that, using Te nanowires as sacrificial templates, the preparation of PtCu nanowires is possible [35]. The synthesis of pentangle like PtCu nanocrystals was performed by co-reduction of $Pt(acac)_2$, $CuCl_2.2H_2O$ with help of glycine in oleyl amine as a solvent at 220°C for 1 h [36]. With the help of KI as a size controlling agent, it was reported different sizes (110-250 nm) of PtCu hierarchical trigonal bipyramid nanoframes (HTBNFs). Moreover, the evolution of HTBNFs was observed through the formation of small monopods, bipods and multipods at early stages followed by the formation of secondary, territory and higher ordered branches including

HTBNFs [37]. By employing epitaxial growth and ascorbic acid (AA) reduction methods, the deposition of isolated Ru atoms and Ru nanoparticles over Pt_3Mn CNCs was reported [38]. The galvanic replacement and one-pot hydrothermal methods were reported to prepare high index facet controlled PtCu and PtPdCu concave nanocubes [39]. It is possible to prepare Pt–Cu nanodendrites by following a two-step co-reduction method, in which a kinetically controlled slow reduction step and, a fast reduction process is carried [40]. The bimetallic concave and dendritic PtCu bimetallic nanocrystals were synthesized via. one-pot co-reduction of H_2PtCl_6 and $Cu(acac)_2$ by oleyl amine and polyvinyl pyrrolidone (PVP) in an autoclave at 180ºC [41]. In the literature, one-step wet chemical approach was reported to synthesize bimetallic alloy 1D ultrafine Pt-Fe nanowires (NWs) with high yield; $Pt(acact)_2$ and Iron(II) acetate ($Fe(Ac)_2$) as metal precursors, glucose and $W(CO)_6$ as reducing agents, N-Hexadecyltrimethylammonium chloride (CTAC) as surfactant and both oleyl amine (OAm) and 1-octadecene (ODE) as solvents [42]. 3D Pt-Pd nanowire networks of different compositions were prepared by employing an efficient wet-chemical route with high yield by using K_2PtCl_4) and ($Pb(acac)_2$) as precursors, citric acid (CA) as reducing agent, polyvinylpyrrolidone (PVP) as capping agent, and ethylene glycol (EG) as solvent and reducing agent [43]. PtCu metallic hollow structures were synthesized using one pot fabrication approach with the help of appropriate precursors (K_2PtCl_6 and $Cu(acac)_2$), surfactant, reductant, solvent (OAm) and co-surfactant (CTAB) [44]. With similar ingredients, cubic nanocages of PtCu3 with a size of 20 nm were also reported.

Without CTAB, mixed polyhedral Pt-Cu nanocrystals were obtained [45]. With change in precursor $Pt(acac)_2$ instead of K_2PtCl_6 and additional surfactant trioctylphosphine oxide (TOPO), PtCu high index concave nanocubes were prepared. In general, the purpose of CTAB and TOPO as coordinating ligands is to establish the co-reduction of the Pt and Cu salt precursors by tuning their redox potentials [46]. The preparation of Pt_3Cu wavy nanowires was performed with the use of PNIPAM-NH_2 as capping agent that preferential adsorbed or capped on {100} or {110} facets and therefore provided the growth along the <111> direction [47]. The amine species such as oleyl amine and the product of DMF and KOH are found to be facet sensitive adsorbed ligands which can promote the growth of the nanostructures into desirable shapes [48]. By establishing strong coordination interaction between metal ions and arginine (a structure directing agent), the preparation of dendritic PtCu bimetallic nanoassemblies of tunable compositions ($PtCu_3$, PtCu) was designed [49]. With appropriate precursors of Pt ($H_2PtCl_6H_2O$) and Pd ($PdCl_2$), bayberry tannin (BT) as reducing agent and surfactant, at 120°C, using autoclave, the PdPt dendritic structure nanoparticles are prepared [50]. Through electrochemical deposition, different shape-controlled nanoparticles (Pt tetrahexaheral, polycrysalline and spherical) are prepared followed by deposition of foreign element over the surface. It was reported that by systematic variation of potential perturbation and time limits, it is possible to tune the size and shape of nanoparticles [51]. Synthesis of ultrathin nanowires of 1.1 nm diameter of PtRu was reported by co-reduction of the appropriate metal precursors in the presence of $W(CO)_6$ and CTAB [52]. During one pot hydrothermal synthesis, with $Pt(acac)_2$ and $Ru(acac)_2$ as precursors, OAm as a solvent, and $W(CO)_6$ as reducing agent, different shape-controlled PtRu nanoparticles were prepared with different structure controlling reagent; PtRu nanowires with DDAC (**Figure 2** a and d) and high resolution of PtRu NW (Figure 2 g), PtRu nanorods with HDBAC (Figure 2 b and e) and PtRu nanocubes with CTAB (Figure 2 c and f) and high resolution of PtRu nanocube Figure 2h and its fast Fourier transform (FFT) pattern (Figure 2i) [53].

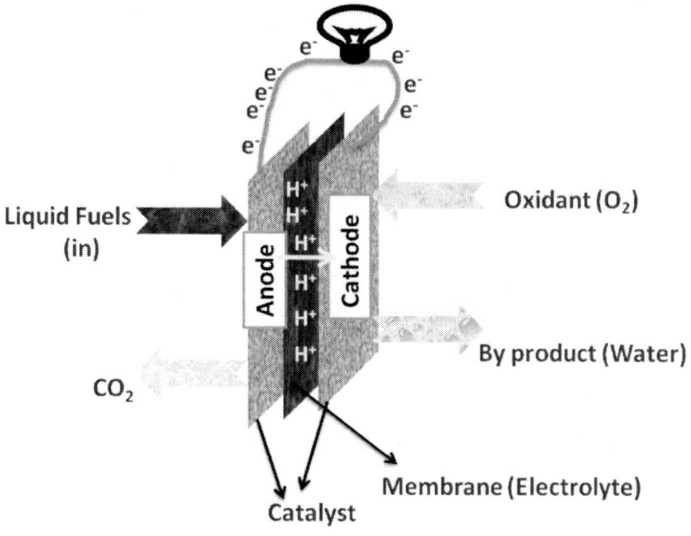

Figure 1. The representation of direct liquid fuel cells (DLFCs) under acidic conditions where in protons exchange occurs through the membrane (electrolyte). This figure is the author's own effort and therefore it does not require any copyright permissions.

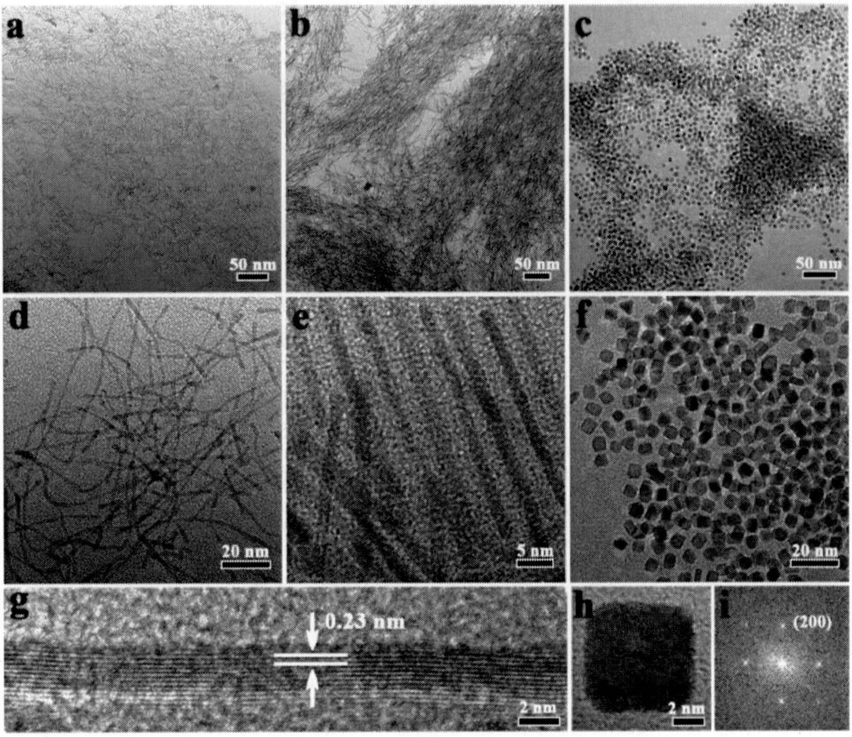

Figure 2. TEM images of (a, d) PtRu NWs, (b, e) PtRu NRs1 and (c, f) PtRu NCs. HRTEM images of an individual (g) PtRu NW and (h) PtRu NC, and the corresponding FFT pattern (i) of (h). Reprinted (adapted) with permission from [53]. Copy right 2018 American Chemical Society.

By implementing galvanic displacement reaction, the preparation of PtRu nanotubes and PtRu over Cu nanowires was reported [54]. Through Stranski-Krastanov growth and galvanic replacement method, not only shape-controlled bimetallic but also multimetallic nanocrystals can be prepared. In this context, PtRuCu nanocrystals of octapods and hexapods were obtained by systematic tuning of reaction time, addition of precursors, aging and temperature. During the preparation of hexapod PtRuCu nanostructures, Cu served as a sacrificial regent to replace with Pt due to the fact that both of them have different reduction potentials; the standard potential of Pt^{+2}/Pt is more positive (1.18 V) than that of Cu^{+2}/Cu (0.34 V) [55]. Moreover, DFT studies suggest that the reduced Cu atoms prefer to deposit on Pt rather than Cu nuclei itself and therefore led to the formation of bimetallic nanocrystals [56]. It was reported that the synthesis of Pt–Ag alloy nano-octahedra by reducing K_2PtCl_4

and $AgNO_3$ precursors with HCHO in aqueous polyallylamine hydrochloride (PAH) solution at 180°C for 4 h is possible. This study found that, the formation of the Pt–Ag alloy nanooctahedra happens only when PAH is present, otherwise irregular and random aggregation nanocrystals are obtained under the standard experimental conditions [57]. It is possible to tune the reduction kinetics of Pt with the help of Zn^{+2} ions resulting in the formation of PtZn bimetallic nanowires of size ~2.2 nm; it was reported that by precisely selecting the solvent and Zn precursor ($ZnCl_2$), the aforementioned shape engineered bimetallic nanoparticles were obtained [58].

3. Performance: Shape Vs. Activity

3.1. Methanol Oxidation Reaction (MOR)

It is important to design an electrocatalyst which provide (a) high open structure with 3D catalyst surface with high surface to volume ratio, and thus effective active metal utilization (b) stepped surface with high index or high energy facets which generally possesses efficient facile electrocatalytic reactions. S. Luo et al., reported self-supported platinum-copper 3D nanoframes of different compositions and different morphologies; concave $PtCu_2$ octopod nanoframes ($PtCu_2$ CONFs) which enclosed with high index facets {553}, {331}, and {221} and ultrathin PtCu octopod nanoframes (PtCu UONFs)) with 12 ultrathin edges. The electrochemical oxidation of methanol with these catalysts suggest that $PtCu_2$ CONFs exhibit better activity and stability compared to commercial Pt/C and PtCu UONFs. It was found that $PtCu_2$ CONFs better activity (7-fold), lower onset potential (~205 mV) and superior CO tolerance than commercial Pt/C, while PtCu UONFs suffer from breakdown [30]. PtCu hollow nanocrystals with several advantages such as low density, materials utilization demonstrated as efficient MOR electrocatalyst with better SAs than commercial Pt black and Pt/C catalysts [59]. To affirm the enhanced performance of hollow structures, novel cubic-type PtCu3 nanocages were demonstrated as efficient and CO-tolerant electrocatalysts for MOR.

The proposed nanocages exhibited better current density and I_f/I_b ratio compared to $PtCu_x$ and commercial Pt black (Johnson Matthey) catalysts [45]. To achieve similar objectives, it was reported that highly open rhombic dodecahedral PtCu nanoframes exhibit improved MOR kinetics under harsh acidic conditions (0.1 M HClO4).

This study observed that PtCu nanoframes exhibited MA of 0.39A/mg_{Pt}, which is 3.9 and 22 times better compared to those of the commercial Pt/C (0.10 A/mg_{Pt}) and the commercial Pt black (0.018 A/mg_{Pt}), respectively and stability upto 200 sweeping cycles [33]. It was reported that hexapod concave Pt/Cu nanoparticles bounded with more atomic steps and defects serve as a better electrocatalyst compared to Pt/Cu nanoparticles; the trend in activity (forward peak current densities) was found to be Pt Black < Pt/Cu NPs < Pt/Cu hexapod concave nanocrystals [60].

In similar lines, due to attributes of unique integrated network within porous structure, high surface area (~68 $m^2.g^{-1}$), and cumulative effect between two metallic components (Pt and Cu), ultrathin dendritic Pt_3Cu triangular pyramid caps are reported as efficient electrocatalysts for methanol oxidation under harsh acidic conditions (0.5 MH_2SO_4+1M MeOH) [61]. Among the PtCu bimetallic ($PtCu_3$, $PtCu_2$, PtCu and Pt_3Cu) concave and dendritic nanocrystals, it was observed that $PtCu_2$ provided the exceptional MOR activity and strong CO poisoning resistance and claimed that the unique concave dendritic morphology of bimetallic counterparts is responsible for better performance [41]. Compared to the shape engineered PtCu based nanocatalysts such as Pt_1Cu_2 nanowires [62], Pt_3Cu icosahedra [32], Pt_3Cu pyramid caps [61], and hollow Pt_1Cu_1 [59], it was reported that Pt_3Cu wavy nanowires of diameter of 2 nm could exhibit the best SA towards MOR. The enhanced performance is attributed to the following aspects: (i) the ultrathin nanowires possess with better exposed active sites due to high surface to volume ratio; (ii) abundant surface defects such as atomic steps and crystal face defects were found around the proximity of crystal domains of nanowires which generally act as active sites for O_2 adsorption, activation and desorption and hence promote efficient oxidation process; (iii) favorable for electron transport and facile transfer of reactants, and products, thus promoting the MOR kinetics [47]. The composition dependent and synergistic aspects were demonstrated with PtCu nanowires of ultrathin diameter (~1 nm) as efficient MOR catalysts. During the study, the composition of $Pt_{32}Cu_{68}$ was found to be optimal to achieve better performance such as MA (2 times), stability, and tolerance to CO poisoning than that of Pt nanowires. It was proposed that, the enhanced performance due to alloying of Cu atoms into Pt lattice may provide CO-maneuvering sites which reduces the impact of poisoning effect of CO on active sites of PtCu nanowires [48]. In fact, nanocrystals with high-index facets possess larger particle sizes, resulting in limited surface area and, consequently, low Pt utilization efficiency or low mass activities. To take the advantages from both high density of high index facets and high surface area, a synthesis approach is carved to produce unique screw thread like Pt-Cu alloy nanowires. These special nanowires of composition $PtCu_{2.1}$ dominant with {331} and {221} facets with high surface area (~47 $m^2.g^{-1}$) performed better than $PtCu_{1.8}$ NPs and Pt/C. The collective features of screw thread like nanocrystals such as 1D structure, alloy effect, Pt rich surface and high index facets promised the advances in the design of electrocatalyst with best MOR activity and stability to CO poisoning [34]. During methanol oxidation under harsh electrochemical conditions, the synergistic aspects between Pd and Pt with improved electrocatalytic performance, CO tolerance and durability was established by the design of PtPd dendrite structures with boundary and corner atoms as active sites. Among the prepared compositions of dendrite structures (($Pd_{82}Pt_{18}$, $Pd_{69}Pt_{31}$, $Pd_{52}Pt_{48}$ and $Pd_{36}Pt_{64}$)), $Pd_{52}Pt_{48}$ exhibited the optimum performance compared to commercial Pt/C (20 wt%) [50]. Liu H. X., et al., studied the importance of catalyst that contains stepped surfaces for electrooxidation of methanol; for a case study, authors performed the study on tetrahexhedrel Pt nanocrystals which are bounded by {730} with some {210}, {310} or {520} facets. It was observed that, Ru modified tetrahexhedrel Pt nanocrystals (THH Pt) outperform commercial PtRu catalysts with superior electrocatalytic currents, CO_2 yields and low (~100 mV) onset potential. To complement the achieved activities, *in-situ* FTIR was performed to understand underlying mechanism of electrooxidation of methanol and found that Ru modification over THH Pt promoted the non-poisoning path through the formation of formic acid [51]. Among the low-index ({100},{111},{110}) facet terminated PtRu nanostructures such as nanocubes, nanowires and nanorods, it was identified that {111} terminated PtRu nanowires exhibited the higher stability (~2.28 times) and MA (~4.32 times) than {100} the PtRu nanocubes and commercial Pt/C. From the DFT calculations, the adsorption energies of CO_{ads} and OH_{ads} on PtRu NWs terminated with {111} facets found to be optimal compared to those on PtRu NCs terminated with {100} facets and thus PtRu NWs contributed to enhanced MOR activity [53]. The preparation of 1D nanostructures such as nanotubes, nanowires, nanorods with exposed active facets could be a promising strategy for the design of efficient anode catalyst for DMFC applications. In this context, PtRu nanotubes and PtRu coated Cu nanowires provided improved kinetics of electrooxidation of methanol compared to commercial carbon supported PtRu. The enhanced SA was attributed to the deposition of Ru that causes the shift in d-band center of Pt which led to weakening of CO (a poisonous intermediate) bonding [54]. It was observed that trimetallic hexapods of PtRuCu exhibited better performance (3.8 fold MA and 8.2 fold SA) than the commercial Pt for the methanol oxidation in the acidic solution. The enhanced performance, durability, and CO tolerance is attributed to the following: (i) The synergistic aspects between Pt, Ru and Cu, (ii) the formation of Ru (OH^-) which promotes the oxidation of CO and (iii) the presence of Cu along with Pt can induce the changes in d-band centre that benefits the weakening of CO bonding with Pt. One of the important index parameter that is the ratio between forward current (I_f) to backward current (I_b) can provide the awareness of tolerance of the catalyst to the poisonous carbonaceous species such as CO, HCHO and HCOOH. During the MOR, the forward anodic scan is generally ascribed to the oxidation of methanol molecules on the catalyst surface, while the back ward scan is assigned to the continuous oxidation of residual or incomplete oxidized carbonaceous intermediates during the forward scan (CO, $HCOO^-$, and HCO^-). The calculated values of I_f/I_b of carbon supported hexapod PtRuCu provided a much higher value (1.32) compared to Pt/C (0.84) which suggests that the aforementioned catalyst exhibits much more stability than the later [55]. Almost similar I_f/I_b results are achieved with Pt_3Cu dendritic structured nanoparticles which are made of dominant (111) facets and coalescence of small dendritic seeds. In this study, along with better MA and SA values, Pt_3Cu dendritic nanoparticles provided better I_f/I_b (1.32) than commercial Pt/C (0.87 V); this implies that Pt-Cu alloy can efficiently oxidize methanol with less prone to poisoning effect of intermediates [40]. Nanoparticles with coverage of high index facets were demonstrated as active MOR catalyst. It was found that PtCu concave nanocubes which are covered with high index facets {511} performed better than commercial Pt/C. The proposed d-band theory suggested that the presence of Pt and Cu could impart small compressive strain due to their lattice mismatch and hence decrease the binding strength of CO on the catalyst surface [46]. Sun X et al., reported for the first time about the importance of bimetallic PtCu nanocrystals which are covered with twin defects for the improved performance of MOR and ORR. They discovered that (111) facets dominant Pt_3Cu icosahedra nanocrystals with twin defects also known as multiple twinned nanocrystals exhibited better activities compared to (111) facets dominant single crystalline Pt_3Cu octahedra without any defects and commercial Pt black; it was demonstrated through CV that recorded in the presence of methanol (0.2 M) in 0.1M $HClO_4$ solution (Figure 3a) and calculated MAs and SAs (Figure 3b). From this study, it was understood that the presence of twin defects and hence lattice strains can accommodate the better opportunity for optimum adsorption, activation, and dissociation capacities for not only O_2 but also small fuel cell molecules such as methanol, ethanol, formic acid and ethyleneglycol [32].

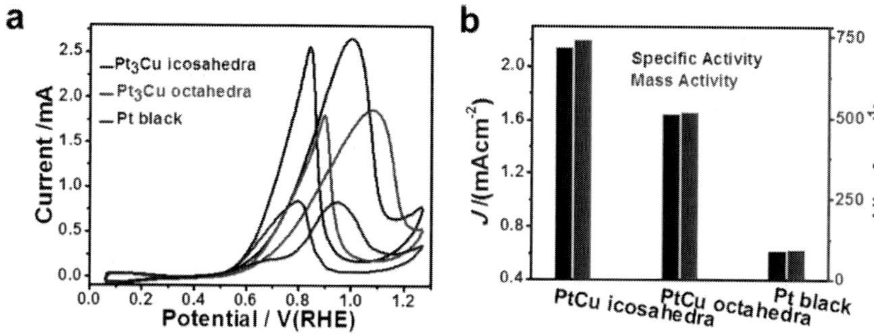

Figure 3. (a) CVs of Pt_3Cu of icosahedra (black), octahedral (red), and Pt black (conditions are oxygen free 0.1M $HClO_4$ and 0.2M CH_3OH solution with a sweep rate of 50 mV s^{-1}). (b) SA and MAs of the three catalysts. Reprinted (adapted) with permission from [32]. Copyright 2015 American Chemical Society.

In support of this observation, A. X. Yin et al., demonstrated the synthesis of PtCu and PtPdCu high index facet (HIFs) concave nanocubes (CNCs) and tested their ability for MOR in acidic solutions. In general, HIFs covered with active sites via several defects such as atomic steps, ledges, and kinks at the surface of catalyst and these sites are phenomenal in contributing to the enhanced performance. Among the catalysts tested, PtPdCu of optimum composition showed the best activity and stability for MOR compared to Pt/C, Pt CNCs, and PtCu CNCs due to shape and synergistic effects [39]. In fact, several experimental observations suggests that elongated Pt nanostructures such as nanorods and nanowires are less prone to dissolution, Ostwald ripening, and aggregation than spherical or polycrystalline nanoparticles. In a study, (111) dominant PtRu nanowires with a diameter of 1.1 nm was employed for the MOR under alkaline (0.1 M KOH) electrolytes and the activity and stability results are compared with commercial Pt/C. It was found that the MA of ultrathin PtRu NWs/C is 3.7 times higher than that of commercial Pt/C. The stability of the ultrathin PtRu NWs/C was outperforming than the commercial Pt/C. From the *in-situ* FTIR results, it was concluded that PtRu counterparts successfully mitigated the poisoning problem of reaction intermediates towards MOR [52]. Similarly, ultrathin PtCu nanowire networks reported as an efficient electrocatalyst for MOR with 3.77 times (MA) and 2.71 times (SA) better compared to commercial Pt/C catalyst due to the fact that it shows unique structure and modified electronic effect [62]. With advantage of bimetallic synergistic effect and unique structural features, it was reported that 3D Pt_3Co nanowire assemblies (NWAs) stands as a promising candidate materials for MOR catalysts compared to commercially available Pt/C, Pt black, Pt/CB (carbon black), PtRu/CB and even Pt NWAs. It was established that, compared to Pt alone, with Pt_3Co catalysts, the presence of Co facilitates the oxygen source which helps in CO oxidation at lower potentials [31]. In order to identify the potential candidate materials for electrochemical energy conversion applications, it was considered that the nanoassemblies of metallic nanocrystals could be promising compared to monodispersed nanocrystals due to the fact that the emergence of collective behavior can provide abundant porosity and thus high internal reactive surface area. Such an attempt was disclosed by the design of dendritic PtCu bimetallic nanoassemblies with tunable composition with improved performance (electrocatalytic activity and stability) than commercial Pt/C [49].

It is convincing to note that the rare use of the bimetallic counterparts of PtAg as catalysts for any applications is due to the difficulty in their preparation attributed to (i) distinct standard reduction potentials leading to different reduction kinetics of Pt^{n+} and Ag^{+} precursors; (ii) significant (4.2%) lattice mismatch between Pt and Ag atoms; and (iii) the large miscibility gap between Pt and Ag in the bulk (e.g., the formation of Pt–Ag alloys is possible with the atomic composition of Ag : Pt < 2 : 98 or Ag : Pt > 95 : 5at 400°C). G. T. Fu et al., reported PtAg nanooctahedra dominant with (111) facets as efficient electrocatalyst for MOR; it was observed that, with PtAg nanooctahedra alloy, the current density during the positive scan and stability ~3 and 4.8 times, respectively higher than that of commercial Pt/C [57].

3.2. Formicacid Oxidation Reaction

For the development of DLFCs, formic acid is also considered a promising liquid fuel due to similar high energy density and renewable nature, comparable to other fuels such as methanol and ethanol. In this context, Pt-Zn bimetallic nanowires are reported to be efficient catalyst with better performance metrics; the onset potential of FAOR is 580 mV vs. SCE which is much lower compared to commercial Pt/C (590 mV) and Pt black (581 mV), while specific (mass) activities achieved with Pt-Zn bimetallic nanowires 7.5 (4.7) times and 6.8 (2) times, respectively better than Pt black and Pt/C counterparts [58]. S. Chen et al., demonstrated the composition variation of PtCu hierarchical trigonal bipyramid nanoframes (HTBNFs) towards FAOR. Due to unique structure and possible synergistic effects between Pt and Cu, it was found that $Pt_{3.32}Cu$ HTBNFs performed 5.5 (SA) and 2.1 (MA) times better than commercial Pt/C [37]. With similar objective of achieving enhanced electrocatalytic oxidation of formic acid, F. Nosheen et al., demonstrated the synthesis of octahedral Pt-Cu nanoframes and provided the better activity than commercial Pt/C and Pt black [63]. While, it was reported that Octahedral $Pt_{25}Cu_{75}$ bimetallic counterparts exhibits superior activity and highest anti-poisoning ability towards FOR [56]. During the cathodic scan of FOR, it was observed that PtCu nanooctahedra showed low activity than that of PtCu and Pt nanocubes. It is interesting to note that, with the decrease in activity to 20% of its initial value, within 20 S with PtCu nanocubes, 40 S with Pt nanocubes and 265 S with PtCu octahedral and thus it is evident that (111) facet dominant PtCu nanoparticles exhibit less activity but more stability than its cubic shape counterparts which dominant with (100) facets [64]. From the experimental and theoretical studies, it is evident that the enhancement in activity, stability and CO anti poising ability with bimetallic and multimetallic nanocrystals compared to its monometallic and commercial Pt/C, Pt black catalysts is due to the effect of composition, surface structures and shape. For instance, in the case of Pt-Ni-Cu alloy of different shapes such as dendritic, crossed and HIFs, provided better electrocatalytic activity towards FOR compared to mono metal and binary metal alloy. The following reasons are cited to convince the better activity and stability: (1) the tuning of surface electronic properties due to alloying with other metals, (2) The addition of Ni and Cu elements drives the down shift of d-band center which reduces $Pt\text{-}CO_{ads}$ bond energy and weakens chemical adsorption of oxygen containing species such as CO_{ads} (3) possessing of high index facets (HIFs) such as atomic steps, ledges and kinks at surface provide more active sites [29]. Apart from synergy of Pt and Ni, the shapes (concave nanocube (CNC) enclosed by {610}, {510}, {410}, and {720} facets and hexoctahedral (HOH) enclosed by {521}, {532} facets) of Pt-Ni alloy nanoparticles contributed to the enhanced electrocatalytic activities (MOR, FOR, and ORR) [65].

3.3. Ethanol Oxidation Reaction

Among the different DLFCs, ethanol based FCs attracts special attention from the research community due to tones of ethanol production as a byproduct from various bio-diesel industries including sugar cane factories. Therefore, it is imperative to use ethanol as a fuel to produce electricity with a FC device at better efficiency and stability. In view of this fact, various shape-engineered bimetallic nanocrystals have been designed and developed for the efficient

DEFCs. It is reported that, among the prepared morphologies of Pd-Cu alloy nanoparticles (from near-spherical to nanorods and nanowire), near spherical Pd-Cu nanoparticles exhibited better activity and stability compared to commercial Pd/C towards EOR. The enhancement in performance with near spherical Pd-Cu nanoparticles is attributed to more edge truncation and synergy of bimetallic nanoparticles [66]. In a different study, PtCu nanowires exhibited better electrocatalytic activity compared to Pt NWs and commercial Pt black catalysts due to negative shift of Pt 4f which led the optimum binding abilities of CH_3CO_{ads} and OH_{ads} [35]. The PtCu alloy with Cu rich {110} facets exhibited superior electrocatalytic performance compared to PtCu hollow structure where in selectively etched Cu. The DFT studies suggest that, it is essential for the presence of OH* species on the surface to detoxify the CO poisoning during the electrocatalysis of target reaction (FOR, MOR, EOR). In the case of PtCu alloy, the presence of Cu could influence the d-band center of Pt which leds to the formation of the key intermediate OH* species from water splitting and hence promote the electrochemical oxidation of liquid fuels [67]. In general, carbon based materials due to features of high surface area, high electrical conductivity and relatively stable under harsh electrochemical conditions are preferred as support materials for noble active metal catalysts during the design of electrocatalysts for energy conversion devices. However, at higher working potentials (>0.7 V), the carbon support is prone to dissolution, aggregation and Ostwald ripening of active metal which causes the poor durability of carbon supported catalyst during fuel cell operation. Therefore, it is crucial to design alternative support materials with better properties which can provide better durability of a fuel cell. In this regard, polyaniline as a conducting polymer attracted significant attention as a support material which is alternative to carbon because of high chemical and electrochemical stability, easy synthesis, environmental friendly, better adhesion and hydrophilic properties. A. L. Wang et al., demonstrated the synthesis of Pd/PANI/Pd sandwich structure nanotube arrays (SNTAs) as efficient electrocatalyst for EOR. The prepared SNTAs exhibited the improved electrocatalytic properties (activity and stability) compared to Pd nanotube arrays and commercial Pd/C; Figure 4 represents the CVs of aforementioned catalysts without (a) and with (b) methanol in oxygen free alkaline solution. The improved performance metrics are attributed to; (i) electron delocalization between Pd d orbitals and PANI π-conjugated ligands provided facile electron transfer from Pd to PANI due to sandwich like structure (ii) fast transport and short diffusion paths for reactants, intermediates and products (iii) better materials utilization [68].

F. Saleem et al., reported the synthesis of ultrathin Pt-Cu nanosheets and nanocones and tested for EOR under acidic conditions. Before testing for EOR, the prepared catalysts were treated under UV light of wavelengths (185 and 254 nm) for 12 h to remove the adsorbed impurities. Both PtCu nanoarchitectures (nanosheets and nanocones) exhibited better performance than commercial Pt black and Pt/C due to unique geometry with exposed surfaces [69]. In order to achieve better EOR kinetics by promoting C-C bond breaking at lower potentials, Sn is used as a bifunctional element with Pt because Sn can provide surface oxygen containing species which are helpful for the removal of poisonous species like CO from the Pt surface. In view of this, it is reported that sub nm PtSn ultrathin nanosheets dominant with highly reactive exposed (111) facets which can exhibit MA with 2.3 and 5.3 times are better than Pt/C and Pt black respectively for EOR [70]. In many catalytic applications, both theoretical and experimental studies suggest that crystal defects containing catalyst enabled better catalytic properties compared to pristine (without defect) of same type and it is expected due to the breakage of electron-hole symmetry. The irradiation strategy was implemented to synthesize PtPb nanoplates enclosed by dislocations, sub grain boundaries and small amorphous domains which expected to maximize the fuel cell catalysis. Defect engineered PtPb nanoplates exhibited better performance towards electrocatalysis of ORR, EOR and MOR [71].

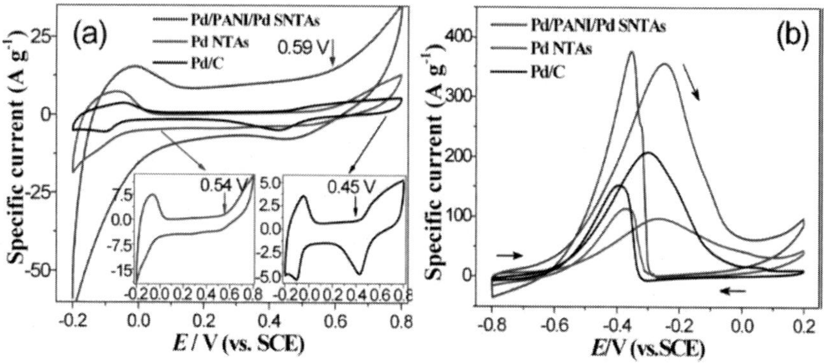

Figure 4. (a) CVs of the Pd/PANI/Pd SNTAs, Pd NTAs, and Pd/C catalysts recorded in O_2 free solution of 1.0 M H_2SO_4. (The inset at right in (a) is the magnified CV of Pd/C catalyst and the inset at left in (a) is the magnified CV of Pd NTAs.) (b) CVs of the Pd/PANI/Pd SNTAs, Pd NTAs, and Pd/C catalysts recorded in O_2 free solution of 1.0 M ethanol + 1.0 M NaOH at 50 mV/s. Reprinted (adapted) with permission from [68]. Copy right 2013 American Chemical Society.

3.4. Ethylene Glyol Oxidation

Ethylene Glycol being known as a typical small organic molecule, attracted community to investigate its electrochemical oxidation reaction kinetics to pave the practical electrochemical energy-conversion devices. In this regard, many efforts were devoted to identify a catalyst with better activity and durability (CO poisoning resistant) and developed many catalysts and reported in the literature. Among them, the present chapter is aiming to inform the knowledge about the effect of specific facets (by shape-control) and bimetallic synergistic aspects towards enhanced EGOR performance. Xu H et al., demonstrated that the pentangle like PtCu nanocrystals of MA and SA can perform ~5 and ~5.1 times better than commercial Pt/C. It was predicted that pentangle like PtCu nanocrystals possesses specific facets which can easily break the chemical bonds during the EGOR [36]. In order to design a stable and efficient electrocatalysts for EGOR, it is imperative to engineer the components of a catalyst to endow with ligand and local strain effects which can trigger the surface electronic and synergistic effects. In this regard, Wang Y et al., demonstrated the effect of doping of isolated Ru atoms (Pt_3Mn-Ru) and Ru nanoparticles (Pt_3Mn@Ru) on to surface layer of Pt_3Mn (concave nanocubes (CNCs)) which posses with high index facets for EGOR. The study found that SA of Pt_3Mn-Ru (1.32 $mAcm^{-2}$) was 1.47, 3.07 times better than Pt_3Mn@Ru (0.9 $mAcm^{-2}$) and pure Pt_3Mn CNCs (0.43 $mAcm^{-2}$). From the *in-situ* FTIR studies, it was uncovered that the C_2/C_1 (C2 species: glycolaldehyde, glyoxal, glycolic acid, glyoxylic acid, oxalic acid; C1 species: formaldehyde, formic acid, CO, CO_2) decreasing with Pt_3Mn-Ru catalyst which suggests that the EGOR follow the pathway of favorable C-C bond cleavage and promote the complete oxidation of EG to CO_2. Moreover, it was identified that the signal associated with CO_L species is weaker which confirms that Pt_3Mn-Ru CNCs is a potential catalyst to remove or promote the better oxidation of the poisonous CO_{ads} species. Furthermore, DFT studies unraveled the reason for high activity and durability of Pt_3Mn-Ru CNCs which is due to low activation barrier (1.69 eV) and an energy-favorable position (2.88 A°) for reaction between CO_{ads} and OH_{ads}. Altogether, it was found that the catalyst possesses defect structure (lattice disorder, gap atoms, and amounts of high density of under coordinated Pt atoms (steps, kinks and edges) and synergistic and strain effects could contribute to the better performance of EGOR [38]. The superior activity and stability with PtFe nanowires was observed for EGOR and is reported to be 3.9 times better performance than commercial Pt/C catalyst. The better performance was attributed to the features of anisotropy, superior flexibility, high surface area and excellent conductivity of PtFe nanowires [42]. As discussed, CO poisoning during the oxidation of alcohol is the culprit for the poor durability of electrocatalyst towards meeting the commercial success of alcohol based fuel cells and therefore it is a perennial task to find the potential candidate materials for efficient and durable alcohol oxidation reaction. One such attempt is, due to fact that high tolerance toward CO poisoning, specific elements such as Pb and Bi alloying with Pt is found to be a promoting effect in alcohol electrooxidation. In this regard, Y. Feng et al., designed 3D based PtPb nanowire networks as electrocatalyst for EGOR. This material exhibited excellent activities in both alkaline and acidic solutions; the MA in acidic solution is 0.73 A.mg_{Pt}^{-1} and in alkaline media, it is 13.5 A mg_{Pt}^{-1} which is 3.3 and 4.8 times better than those of commercial Pt/C, respectively. The I_f/I_b of these catalysts show that, $Pt_{4.5}Pb$ NWs (1.59) exhibit higher than those of $Pt_{5.7}Pb$ NWs (1.22), Pt NWs (1.08), and Pt/C (1.09), suggested their high antipoisoning property. The highlight of this work is the durability of catalyst and it was demonstrated that, after 1500 sweeping cycles, the networked $Pt_{4.5}Pb$ NWs show the 70% of the initial catalytic activity, which is better than those of the other catalysts $Pt_{5.7}Pb$ NWs (45%), Pt NWs (30%), and Pt/C (21%). The enhanced performance of networked $Pt_{4.5}Pb$ NWs was attributed to their unique 3D structure with large specific surface area, hierarchical porosity and high density of defects and steps [43].

Conclusion

The literature suggests that, under controlled experimental conditions, an array of bimetallic alloys, well-crafted in terms of size, shape, and composition, can be elegantly synthesized. The performance metrics such as onset potential, over potential, mass activity, specific activity, and stability of shape-controlled bimetallic alloys outperform than that of monometallic and commercial catalysts (Pt black and Pt/C). It was identified that the formation of intermediate product (CO) hinders the stability of the catalyst and hence results in inefficient performance of DLFCs. This issue is successfully mitigated through the design of shape-controlled bimetallic nanocrystals, exhibiting improved resistance to CO poisoning and consequently attaining enhanced performance. In general, the improved performance (activity and stability) of shape-controlled bimetallic catalysts towards the electrochemical oxidation of LFs is attributed to the following: (i) the exposed active facets are easily accessible to reactants (ii) better surface/volume ratios through 1D (nanowires, nanorods), 2D (nanosheets, nanoplates) type architectures (iii) due to synergy of elements, optimum d-band center position is achieved which is phenomenal in contributing to not only faster kinetics but also to stability by promoting the formation of active intermediate species on the surface (for instance; OH species), (iv) high contact area between the support and active metal through the design of sub nanometer thin nanosheets, wires, and rods, (v) some nanoarchictures such as nanoframes, hollow structures, etc. possesses with various type of defect sites (atomic steps, kinks, vacancies, ledges) which are contributing towards efficient kinetics. Indeed, shape-controlled bimetallic alloy catalysts face challenges in terms of scalability, with at least a gram-scale production expected for industrial applications. The literature suggests that the synthesis of the material is currently achieved at low yields (in milligrams). Therefore, efforts are required to scale up the preparation of materials to the gram scale without compromising the performance of the catalyst. Though the shape-controlled bimetallic alloy shows promising activity, still durability for longer hours remains elusive. Most of the literature is replete with half cell reaction towards identifying the performance of the catalyst, efforts are therefore needed to test at realistic fuel cell conditions; perhaps, as mentioned earlier, the low yield (mg) of synthesis of materials is the hindering fact. It is understood that CO_2 is one of the byproducts during the electrochemical oxidation of LFs, and as a greenhouse gas, it might contribute to the carbon footprint. Perhaps, the integrated CO_2 electrolyzer technology, which enables the reduction of CO_2 to useful chemicals (e.g., methanol, ethanol), could provide a possible solution for mitigating carbon emissions. *In-situ* electrochemical studies which simulate the real fuel cell operation conditions might provide the better knowledge on the formation of intermediate species and their interactions with catalyst surface which is imperative to design and development of better catalyst. Due to the exceptional properties of LFs, it is possible to offer alternative solutions through further advancements in DLFCs as energy conversion devices for both portable and automobile applications. This might replace conventional fuels (petrol, diesel, natural gas) in the future.

References

[1] Umberto, L. (2014). Overview on Fuel Cells, *Renewable Sustainable Energy Rev.*, 30, 164–169.

[2] Sazali, N., Salleh, W. N. W., Jamaludin, A. S., and Razali, M. N. M. (2020). New Perspectives on Fuel Cell Technology: A Brief Review, *Membranes.*, 10, 1-18.

[3] Borup, R., Meyers, J., Pivovar, B., Kim Y. S., Mukundan R., Garland, N., Myers, D., Wilson, M., Garzon, F., Wood, D., Zelenay, P., More, K., Stroh, K., Zawodzinski, T., Boncella, J., McGrath, J. E., Inaba, M., Miyatake, K., Hori, M., Ota, K., Ogumi, Z., Miyata, S., Nishikata, A., Siroma Z., Uchimoto, Y., Yasuda, K., Kimijima, K., and Iwashita, N. (2007). Scientific Aspects of Polymer Electrolyte Fuel Cell Durability and Degradation, *Chem. Rev.*, 107, 3904-3951.

[4] Sharaf, O. Z., and Orhan, M. F. (2014). An over view of fuelcell technology: Fundamentals and applications, *Renewable Sustainable Energy Rev.*, 32, 810–853.

[5] Castelvecchi, D. (2022). The Hydrogen Revolution, *Nature.*, 611, 440-443.

[6] Cimenti, M., and Hill, J. M. (2009). Direct Utilization of Liquid Fuels in SOFC for Portable Applications: Challenges for the Selection of Alternative Anodes, *Energies.*, 2, 377-410.

[7] Soloveichik, G. L. (2014). Liquid Fuelcells, *Beilstein J. Nanotechnol.*, 5, 1399–1418.

[8] Ong, B. C., Kamarudin, S. K., and Basri, S. (2017). Direct Liquid Fuelcells: A review, *Int. J. Hydrog. Energy.*, 42, 10142-10157.

[9] Shaari, N., Kamarudin, S. K., Bahru, R., Osman, S. H., and Ishak, N. A. I. M. (2021). Progress and Challenges: Review for Direct Liquid Fuel Cell, *Int J Energy Res.*, 45, 6644–6688.

[10] Karim, N. A., and Kamarudin, S. K. (2012). An overview on non-platinum cathode catalysts for direct methanol fuel cell, *Appl. Energy.*, 103, 212-220.

[11] Ozoemena, K. I. (2016). Nanostructured Platinum-Free Electrocatalysts in Alkaline Direct Alcohol Fuel Cells: Catalyst Design, Principles and Applications, *RSC Adv.*, 6, 89523–89550.

[12] Antolini, E., and Gonzalez, E. R. J. (2010). Alakaline Direct Alcohol Fuel Cells, *Power Sources.*, 195, 3431–3450.

[13] Zhao, X., Yin, M., L., Liang, L., Liu, C., Liao, J., Luc, T., and Xing, W. (2011). Recent Advances in Catalysts for Direct Methanol Fuel Cells, *Energy Environ. Sci.*, 4, 2736-2753.

[14] Kakati, N., Maiti, J., Lee, S. H., Jee, S. H., Viswanathan, B., and Yoon, Y. S. (2014). Anode Catalysts for Direct Methanol Fuel Cells in Acidic Media: Do We Have Any Alternative for Pt or Pt-Ru? *Chem. Rev.*, 114, 12397−12429.

[15] Glüsen, A., Dionigi, F., Paciok, P., Heggen, M., Müller, M., Gan, L., Strasser, P., Dunin-Borkowski, R. E., and Stolten, D. (2019). Dealloyed PtNi-Core–Shell Nanocatalysts Enable Significant Lowering of Pt Electrode Content in Direct Methanol Fuel Cells, *ACS Catal.*, 9, 3764−3772.

[16] Rizo, R., and Cuenya, B. R. (2019). Shape-Controlled Nanoparticles as Anodic Catalysts in Low-Temperature Fuel Cells, *ACS Energy Lett.*, 4, 1484−1495.

[17] Porter, N. S., Wu, H., Quan, Z., and Fang, J. (2013). Shape-Control and Electrocatalytic Activity-Enhancement of Pt-Based Bimetallic Nanocrystals, *Acc. Chem. Res.*, 46, 1867–1877.

[18] Lai, S. C. S., Lebedeva, N. P., Housmans, T. H. M., and Koper, M. T. M. (2007). Mechanisms of Carbon Monoxide and Methanol Oxidation at Single-Crystal Electrodes, *Top Catal.*, 46, 320–333.

[19] Tripkovic, A. V., Popovic´, K. Dj., and Lovic, J. D. (2001). The Influence of the Oxygen-Containing Species on the Electrooxidation of the C1–C4 Alcohols at Some Platinum Single Crystal Surfaces in Alkaline Solution, *Electrochim. Acta*, 46, 3163–3173.

[20] Solla-Gullo´n, J., Vidal-Iglesias, F. J., Lo´pez-Cudero, A., Garnier, E., Feliu, J. M., and Aldaz, A. (2008). Shape-Dependent Electrocatalysis: Methanol and Formic Acid Electrooxidation on Preferentially Oriented Pt Nanoparticles, *Phys. Chem. Chem. Phys.*, 10, 3689–3698.

[21] Devivaraprasad, R., Nalajala, N., Bera, B., and Neergat, M. (2019). Electrocatalysis of Oxygen Reduction Reaction on Shape-Controlled Pt and Pd Nanoparticles—Importance of Surface Cleanliness and Reconstruction, *Front. Chem.*, 7, 1-25.

[22] Bhol, P., Bhavya, M. B., Swain, S., Saxena, M., and Samal, A. K. (2020). Modern Chemical Routes for the Controlled Synthesis of Anisotropic Bimetallic Nanostructures and Their Application in Catalysis, *Front. Chem.*, 8, 1-26.

[23] Li, G., Zhang, W., Luo, N., Xue, Z., Hu, Q., Zeng, W., and Xu, J. (2021). Bimetallic Nanocrystals: Structure, Controllable Synthesis and Applications in Catalysis, Energy and Sensing, *Nanomater.*, 11, 1-37.

[24] Xia, Y., Xiong, Y., Lim, B., and Skrabalak, S. E. (2009). Shape-Controlled Synthesis of Metal Nanocrystals: Simple Chemistry Meets Complex Physics? *Angew Chem. Inter. Ed.*, 48, 60-103.

[25] Xia, Y., Xia, X., and Peng, H. C. (2015). Shape-Controlled Synthesis of Colloidal Metal Nanocrystals: Thermodynamic versus Kinetic Products, *J. Am. Chem. Soc.*, 137, 7947−7966.

[26] Nalajala, N., Chakraborty, A., Bera, B., and Neergat, M. (2016). Chloride (Cl(-)) Ion-Mediated Shape Control of Palladium Nanoparticles, *Nanotechnology*, 27, 065603 (12pp).

[27] Naresh, N., Wasim, F. G. S., Ladewig, B. P., and Neergat, M. (2013). Removal of Surfactant and Capping Agent from Pd Nanocubes (Pd-NCs) Using Tert-Butylamine: Its Effect on Electrochemical Characteristics, *J. Mater. Chem. A*, 1, 8553-8559.

[28] Naresh, N., Wasim, F. G. S., Ladewig, B. P., and Neergat, M. (2014). Sodium Borohydride Treatment: A Simple and Effective Process for the Removal of Stabilizer and Capping Agents from Shape-Controlled Palladium Nanoparticles, *Chem. Comm.*, 50, 9365-9368.

[29] Zhang, P., Dai, X., Zhang, X., Chen, Z., Yang, Y., Sun, H., Wang, X., Wang, H., Wang, M., Su, H., Li, D., Li, X., and Qin, Y. (2015). One-Pot Synthesis of Ternary Pt–Ni–Cu Nanocrystals with High Catalytic Performance, *Chem. Mater.*, 27, 6402–6410.

[30] Luo, S., and Shen, P. K. (2017). Concave Platinum-Copper Octopod Nanoframes Bounded with Multiple High-Index Facets for Efficient Electrooxidation Catalysis, *ACS Nano.*, 11, 11946−11953.

[31] Xia, B. Y., Wu, H. B., Li, N., Yan, Y,; Lou, W. X., and Wang, X. (2015). One-Pot Synthesis of Pt–Co Alloy Nanowire Assemblies with Tunable Composition and Enhanced Electrocatalytic Properties, *Angew. Chem. Int. Ed.*, 54, 1 – 6.

[32] Sun, X., Jiang, K., Zhang, N., Guo, S., and Huang, X. (2015). Crystalline Control of {111} Bounded Pt_3Cu Nanocrystals: Multiply-Twinned Pt3Cu Icosahedra with Enhanced Electrocatalytic Properties, *ACS Nano.*, 9, 7634–7640.

[33] Ding, J., Zhu, X., Bu, L., Yao, J., Guo, J., Guo, S., and Huang X. (2015). Highly Open Rhombic Dodecahedral PtCu Nanoframes, *Chem. Commun.*, 51, 9722-9725.

[34] Zhang, N., Bu, L., Guo, S., Guo, J., and Huang, X. (2016). Screw Thread-Like Platinum-Copper Nanowires Bounded with High-Index Facets for Efficient Electrocatalysis, *Nano Lett.*, 16, 8, 5037–5043.

[35] Hong, W., Wang, J., and Wang, E. (2015). Facile Synthesis of PtCu Nanowires with Enhanced Electrocatalytic Activity, *Nano Res.*, 8, 2308-2316.

[36] Xu, H., Liu, C., Song, P., Wang, J., Gao, F., Zhang, Y., Shiraishi, Y., Di, J., and Du, Y. (2018). Ethylene Glycol Electrooxidation Based on Pentangle-Like PtCu Nanocatalysts, *Chem. Asian J.*, 13, 626-630.

[37] Chen, S., Su, H., Wang, Y., Wu, W., and Zeng, J. (2014). Size-Controlled Synthesis of Platinum–Copper Hierarchical Trigonal Bipyramid Nanoframes, *Angew. Chem. Int. Ed.*, 53, 1 – 7.

[38] Wang, Y., Zheng, M., Sun, H., Zhanga, X., Luan, C., Li, Y., Zhao, L., Zhao, H., Dai, X., Ye, J. Y., Wang, H., and Sun, S. G. (2019). Catalytic Ru Containing Pt3Mn Nanocrystals Enclosed with High-Indexed Facets: Surface Alloyed Ru Makes Pt More Active than Ru Particles for Ethylene Glycol Oxidation, *Appl. Catal. B: Environmental.*, 253, 11–20.

[39] Yin, A. X., Min, X. Q., Zhu, W., Liu, W. C., Zhang, Y. W., and Yan, C. H. (2012). Pt-Cu and Pt-Pd-Cu Concave Nanocubes with High-Index Facets and Superior Electrocatalytic Activity, *Chem. Eur. J.*, 18, 777 – 782.

[40] Taylor, E., Chen, S., Tao, J., Wu, L., Zhu, Y., and Chen, J. (2013). Synthesis of Pt-Cu Nanodendrites through Controlled Reduction Kinetics for Enhanced Methanol Electro-Oxidation, *ChemSusChem*, 6, 1863-1867.

[41] Wang, Y. X., Zhou, H. J., Sun, P. C., and Chen, T. H. (2014). Exceptional Methanol Electro-Oxidation Activity by Bimetallic Concave and Dendritic Pt–Cu Nanocrystals Catalysts, *J. Power Sources.*, 245, 663-670.

[42] Song, P., Xu, H., Wang, J., Zhang, Y., Gao, F., Guo, J., Shiraishi, Y., and Du, Y. (2018). 1D Alloy Ultrafine Pt-Fe Nanowires as Efficient Electrocatalysts for Alcohol Electrooxidation in Alkaline Media, *Nanoscale*, 10, 16468-16473.

[43] Feng, Y., Bu, L., Guo, S., Guo, J., and Huang, X. (2016). 3D Platinum-Lead Nanowire Networks as Highly Efficient Ethylene Glycol Oxidation Electrocatalysts, *Small*, 12, 4464-4470.

[44] Yu, X., Wang, D., Peng, Q., and Li, Y. (2011). High Performance Electrocatalyst: Pt–Cu Hollow Nanocrystals, *Chem. Commun.*, 47, 8094–8096.

[45] Xia, B. Y., Wu, H. B., Wang, X., and Lou, X. W. (2012). One-Pot Synthesis of Cubic PtCu3 Nanocages with Enhanced Electrocatalytic Activity for the Methanol Oxidation Reaction, *J. Am. Chem. Soc.*, 134, 13934–13937.

[46] Qi, Y., Bian, T., Choi, S-I., Jiang, Y., Jin, C., Fu, M., Zhang, H., and Yang, D. (2014). Kinetically Controlled Synthesis of Pt–Cu Alloy Concave Nanocubes with High-Index Facets for Methanol Electro-Oxidation, *Chem. Commun.*, 50, 560-562.

[47] Fu, G., Yan, X., Cui, Z., Sun, D., Xu, L., Tang, Y., Goodenough, J. B., and Lee, J. M. (2016). Catalytic Activities for Methanol Oxidation on Ultrathin $CuPt_3$ Wavy Nanowires With/Without Smart Polymer, *Chem. Sci*., 7, 5414–5420.

[48] Liao, Y., Yu, G., Zhang, Y., Guo, T., Chang, F., and Zhong, C. J. (2016). Composition-Tunable PtCu Alloy Nanowires and Electrocatalytic Synergy for Methanol Oxidation Reaction, *J. Phys. Chem. C*, 120, 10476–10484.

[49] Fu, G., Liu, H., You, N., Wu, J., Sun, D., Xu, Lin., Tang, Y., and Chen, Y. (2016). Dendritic Platinum–Copper Bimetallic Nanoassemblies with Tunable Composition and Structure: Arginine-Driven Self-Assembly and Enhanced Electrocatalytic Activity, *Nano Res*., 9, 755-765.

[50] Zhang, Y., Zhang, J., Chen, Z., Liu, Y., Zhang, M., Han, X., Zhong, C., Hu, W., and Deng, Y. (2018). One-Step Synthesis of the PdPt Bimetallic Nanodendrites with Controllable Composition for Methanol Oxidation Reaction, *Sci China Mater.*, 61, 697–706.

[51] Liu, H. X., Tian, N., Brandon, M. P., Zhou, Z. Y., Lin, J. L., Hardacre, C., Lin, W. F., and Sun, S. G. (2012). Tetrahexahedral Pt Nanocrystal Catalysts Decorated with Ru Adatoms and Their Enhanced Activity in Methanol Electrooxidation, *ACS Catal.*, 2, 708−715.

[52] Yin, K., Chao, Y., Zeng, L., Li, M., Liu, F., Guo, S., and Li, H. (2022). Ultrathin PtRu Nanowires as Efficient and Stable Electrocatalyst for Liquid Fuel Oxidation Reactions, *Energy Mat. Advances*., 2022, 9871842, 8 pages.

[53] Huang, L., Zhang, X., Wang, Q., Han, Y., Fang, Y., and Dong, S. (2018). Shape-Control of Pt-Ru Nanocrystals: Tuning Surface Structure for Enhanced Electrocatalytic Methanol Oxidation, *J. Am. Chem. Soc.*, 140, 1142–1147.

[54] Zheng, J., Cullen, D. A., Forest, R. V., Wittkopf, J. A., Zhuang, Z., Sheng, W., Chen, J. G., and Yan, Y. (2015). Platinum–Ruthenium Nanotubes and Platinum–Ruthenium Coated Copper Nanowires as Efficient Catalysts for Electro-Oxidation of Methanol, *ACS Catal.*, 5, 1468–1474.

[55] Xue, S., Deng, W., Yang, F., Yang, J., Amiinu, I. S., He, D., Tang, H., and Mu, S. (2018). Hexapod PtRuCu Nanocrystalline Alloy for Highly Efficient and Stable Methanol Oxidation, *ACS Catal.*, 8, 7578–7584.

[56] Jia, Y., Su, J., Chen, Z., Tan, K., Chen, Q., Cao, Z., Jiang, Y., Xie, Z., and Zheng, L. (2015). Composition-Tunable Synthesis of Pt–Cu Octahedral Alloy Nanocrystals from PtCu to $PtCu_3$ via Underpotential-Deposition-Like Process and Their Electro-Catalytic Properties, *RSC Adv.*, 5, 18153-18158.

[57] Fu, G. T., Ma, R. G., Gao, X. Q., Chen, Y., Tang, Y. W., Lub, T. H., and Lee, J. M. (2014). Hydrothermal Synthesis of Pt-Ag Alloy Nano-Octahedra and Their Enhanced Electrocatalytic Activity for the Methanol Oxidation Reaction, *Nanoscale*, 6, 12310–12314.

[58] Pei, J., Mao, J., Liang, X., Zhuang, Z., Chen, C., Peng, Q., Wang, D., and Li, Y. (2018). *ACS Sustainable Chem. Eng.* Ultrathin Pt–Zn Nanowires: High-

Performance Catalysts for Electrooxidation of Methanol and Formic Acid, 6, 77–81.

[59] Yu, X., Wang, D., Peng, Q., and Li, Y. (2011). High Performance Electrocatalyst: Pt–Cu Hollow Nanocrystals, *Chem. Commun.*, 47, 8094–8096.

[60] Liu, X., Wang, W., Li, H., Li, L., Zhou, G., Yu, R., Wang, D., and Li, Y., (2013). One-Pot Protocol for Bimetallic Pt/Cu Hexapod Concave Nanocrystals with Enhanced Electrocatalytic Activity, *Scientific Reports*, 1404, 1-5.

[61] Kuang, Y., Cai, Z., Zhang, Y., He, D., Yan, X., Bi, Y., Li, Y., Li, Z., and Sun, X. (2014). Ultrathin Dendritic Pt_3Cu Triangular Pyramid Caps with Enhanced Electrocatalytic Activity, *ACS Appl. Mater. Interfaces.*, 6, 17748–17752.

[62] Xiao, M., Li, S., Zhao, X., Zhu, J., Yin, M., Liu, C., and Xing, W. (2014). Enhanced Catalytic Performance of Composition-Tunable PtCu Nanowire Networks for Methanol Electrooxidation, *ChemCatChem.*, 6, 2825 – 2831.

[63] Nosheen, F., Zhang, Z-C., Zhuang, J., and Wang, X. (2013). One-Pot Fabrication of Single-Crystalline Octahedral Pt-Cu Nanoframes and Their Enhanced Electrocatalytic Activity, *Nanoscale*, 5, 3660-3663.

[64] Zhang, J., Yang, H., Martens, B., Luo, Z., Xu, D., Wang, Y., Zou, S., and Fang, J. (2012). Pt–Cu nanoctahedra: synthesis and comparative study with nanocubes on their electrochemical catalytic performance, *Chem. Sci.*, 3, 3302.

[65] Xu, X., Zhang, X., Sun, H., Yang, Y., Dai, X., Gao, J., Li, X., Zhang, P., Wang, H. H., Yu, N. F., and Sun, S. G. (2014). Synthesis of Pt–Ni Alloy Nanocrystals with High-Index Facets and Enhanced Electrocatalytic Properties, *Angew. Chem.*, 126, 12730 –12735.

[66] Wang, H., Yang, W., Zhang, Q., and Qiu, Q. (2016). Shape-Controlled Synthesis of Palladium-Copper Nanoalloys with Improved Catalytic Activity for Ethanol Electrooxidation, *Inter. J. Electro.*, 2016, 1-8.

[67] Huang, L., Zhang, W., Li, P., Song, Y., Sheng, H., Du, Y., Wang, Y. G., Wu, Y., Hong, X., Ding, Y., Yuan, X., and Zhu, M. (2019). Exposing Cu-Rich {110} Active Facets in PtCu Nanostars for Boosting Electrochemical Performance toward Multiple Liquid Fuels Electrooxidation, *Nano res.*, 12, 1147-1153.

[68] Wang, A. L., Xu, H., Feng, J. X., Ding, L. X., Tong, Y. X., and Li, G. R. (2013). Design of Pd/PANI/Pd Sandwich-Structured Nanotube Array Catalysts with Special Shape Effects and Synergistic Effects for Ethanol Electrooxidation, *J. Am. Chem. Soc.*, 135, 29, 10703–10709.

[69] Saleem, F., Zhang, Z., Xu, B., Xu, X., He, P., and Wang, X. (2013). Ultrathin Pt-Cu Nanosheets and Nanocones, *J. Am. Chem. Soc.*, 135, 18304−18307.

[70] Chen, J. Y., Lim, S. C., Kuo, C. H., and Tuan, H. Y. (2019). Sub-1 Nm PtSn Ultrathin Sheet as an Extraordinary Electrocatalyst for Methanol and Ethanol Oxidation Reactions, *J. Colloid Interface Sci.*, 545, 54–62.

[71] Sun, Y., Liang, Y., Luo, M., Lv, F., Qin, Y., Wang, L., Xu, C., Fu, E., and Guo, S. (2018). Defects and Interfaces on PtPb Nanoplates Boost Fuel Cell Electrocatalysis, *Small.*, 14, 1702259 (1-8).

Chapter 2

Carbon Dioxide Hydrogenation to Methanol: New Insights into the Cu-Based Bimetallic Catalysts

Prabhakar Reddy Kasala[1,*]
Surendar Tonda[2]
and Srikanth Reddy Tulsani[3]
[1]Department of Chemistry and Biochemistry, University of South Carolina, Columbia, South Carolina, USA
[2]School of Architecture, Civil, Environmental and Energy Engineering, Kyungpook National University, Daegu, Republic of Korea
[3]Department of Physics, Malla Reddy College of Engineering and Technology, Hyderabad, Telangana, India

Abstract

CO_2 hydrogenation to methanol has become an attractive field of study in recent years as it transforms the greenhouse gas (CO_2) into value-added chemicals. Copper-based catalysts have been used for this reaction due to their low cost and good selectivity for methanol over other metals. However, the CO_2 conversion efficiency of Cu-based catalysts remains low even at very high reaction pressures. As a result, significant efforts have been directed towards tailoring the Cu-based catalysts to boost CO_2 conversion without reducing methanol selectivity. One effective and practical strategy is to introduce a second metal, particularly Zn, to obtain bimetallic Cu-based catalysts. The addition of Zn to Cu results in a synergistic interaction between Cu and Zn, improving CO_2 activation and

* Corresponding Author's Email: mr.prabhakasala@gmail.com.

In: Bimetals
Editors: Pranjal Saikia, Jayanta Kumar Sarmah and Parikshit GogoiName
ISBN: 979-8-89113-496-6

methanol selectivity. The current chapter provides a detailed discussion on the fundamentals of CO_2 chemistry, the evolution from mono-metallic to bimetallic catalysts for CO_2 hydrogenation, operando studies, and the significance of bimetallic catalysts in CO_2 chemisorption and methanol selectivity. In addition to that, special emphasis has been provided on the status of the reaction and prospects for material design to improve methanol selectivity. Finally, a future outlook for the development of Cu-based bimetallic catalysts is provided.

1. Introduction

The ever-increasing anthropogenic carbon dioxide (CO_2) emissions and the depletion of fossil fuels have sparked renewed interest in CO_2 chemistry [1]. The carbon neutral approach is one of the key aspects for recycling CO_2 through activation and generating valuable feedstocks, and catalysis is a key science to overcome these challenges. In this process, we need a green, efficient, affordable catalytic process to recycle CO_2 and to address the highly growing future energy demands. Among these, selective hydrogenation of CO_2 to methanol (MeOH) is a promising prospect since it enables CO_2 recycling and potential fuel replacements [2]. However, the direct production of methanol is a challenging task at the commercial scale owing to its high thermodynamic stability over the viable products [3]. The CO_2 molecule has linear and symmetric nature with $C = O$ bonds making it thermodynamically more stable. Hence, the identification of a suitable and scalable catalyst is a vital challenge for the sustainable production of hydrocarbons. Thus, the objective of this research is to displace the usual method of methanol production via the syngas route, which has been actively carried out to date using Cu-based catalysts as the primary active component for CO_2 reduction [4]. Methanol production with the syn-gas route has two-step process and low process efficiency limits the industrial scale [5]. Whereas direct conversion of CO_2 to methanol suffers from a low equilibrium CO_2 conversion as the hydrogenation reaction is exothermic in nature. Subsequently, a lower reaction temperature is more promoted towards the MeOH formation. As a result, the design of a catalyst with high CO_2 conversion and high methanol formation is of great interest in catalysis. Carbon dioxide conversion into direct fuels and value-added chemicals has been vigorously pursued for the generation of renewable and sustainable green energy, and significant progress has been made in the discovery of novel, highly active catalysts [6, 7]. It can be proceeded through electrocatalysis, thermal catalysis and photocatalysis [8-

10]. Among these, thermal catalysis gains considerable attention owing to its fast reaction kinetics and flexible combination of active components.

Recently, bimetallic nano catalysts have demonstrated intriguing catalytic trends due to their unique properties, resulting in the combination of the constituent metals [6, 11-13]. Often, bimetallic surface has displayed improved catalytic CO_2 hydrogenation on account of strong synergistic interaction or alloy formation, which promotes the electronic or geometric changes to their counterparts [14]. The surface structure of the d-electron bands and their electron filling may change significantly when a bimetallic alloy is formed due to changes in electronic properties caused by heteroatomic bond formation and geometric irregularity. Heteroatomic bonds results in alteration of the electronic configuration of primary metal consequently changing its chemical properties also known as the electronic/ligand effect. In addition, the geometric irregularity arising from the alloying of two different-sized metals causes lattice distortion, which leads to variations in the bond angle and bond length of the constituent metal. This is referred to as the geometrical (or ensemble) effect results in the changes of constituent metal electronic properties. As shown in Figure 1, bimetallic nanomaterials classified into three types depends upon the configuration of constituent metals and its ordering [15]. a). Intermetallic alloy b). Core-shell structure, c) Solid solution b) Heterostructure.

The heterogeneous catalysts for CO_2 hydrogenation have progressed steadily because of advances in nanomaterial synthesis and operando techniques, notably on Cu-based catalysts [16-18]. This process is still receiving attention because heterogeneous catalysts outperform over homogeneous catalysts in terms of reusability, stability, and ease of handling, allowing for large-scale productions at lower costs. Extensive efforts have been made over the years to produce active and stable copper-based heterogeneous catalysts for CO_2 hydrogenation to methanol and value-added compounds. This chapter focuses on fundamental CO_2 chemistry, surface science investigations pertaining to CO_2 chemistry, operando studies, and the significance of bimetallic catalysts in CO_2 chemisorption and methanol selectivity. Also the present status of the reaction and prospects towards material design to improve the selectivity of methanol is elaborated in this chapter. The literature reveals that bimetallic catalysts supported on various metal oxides are effective catalysts for CO_2 hydrogenation to methanol production. This chapter is divided into the following sections which contains the discussion on thermodynamic and kinetic consideration of CO_2 chemistry, Cu-based monometallic catalysts, bimetallic catalysts, thermodynamic and kinetic considerations, reaction pathways and reaction intermediates, bimetallic catalysts synthesis and finally with future outlook and conclusions.

2. Thermodynamics and Kinetic Consideration

The catalytic performance of CO_2 hydrogenation to a desirable product will be determined by the reaction thermodynamics. Thus, reaction pressure and temperature impact the maximum CO_2 conversion and are defined by thermodynamic equilibrium. The CO_2 hydrogenation to methanol reaction often proceeds by two competing pathways [7]. The first pathway is the direct production of methanol.

$$CO_2 + 3H_2 \rightarrow CH_3OH + H_2O.\ \Delta H^0_{298K} = -49.5\ kJ/mol \tag{1}$$

As an exothermic process (Eq. 1), the direct conversion of CO_2 to methanol is preferred at low reaction temperatures and high pressures. However, slow reaction kinetics and stability of CO_2 molecule leads to high reaction temperature required.

Second, the indirect production of methanol occurs through the catalytic hydrogenation of CO, which is produced by the RWGS process.

$$CO_2 + H_2 \rightarrow CO + H_2O.\ \Delta H^0\ _{298K} = 41.2kJ/mol \tag{2}$$

$$CO + 2H_2 \rightarrow CH_3OH.\ \Delta H^0\ _{298K} = -90.6kJ/mol \tag{3}$$

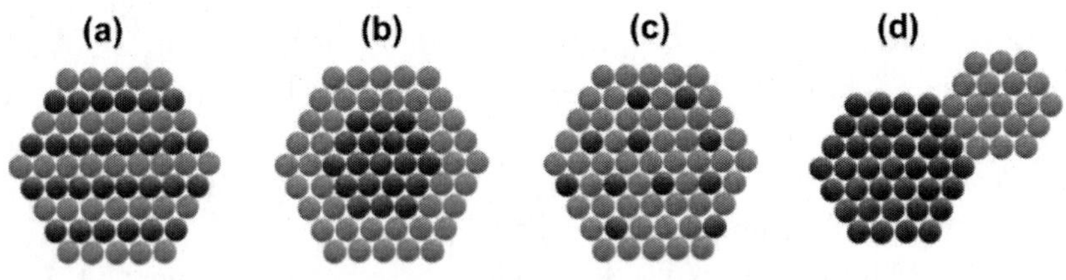

Figure 1. Bimetallic nanostructures: (a). Intermetallic alloy, (b). Core-shell structure, (c) Solid Solution, (b) Heterostructure.

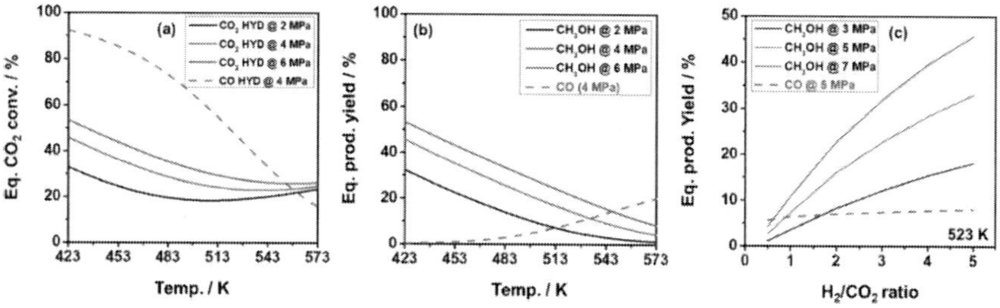

Figure 2. (a) Equilibrium CO_2/CO conversion and (b, c) product yield as a function of temperature, pressure (CO_2/H_2 = 1:3), and H_2/CO_2 ratio. Adapted with permission from reference [7]. Copy right 2020 American Chemical Society.

RWGS process is endothermic and advantageous at high temperatures and low pressures producing more CO and water. In addition, the formation of water molecules during this process deactivates the catalyst. These characteristics influence catalyst performance and process efficiency, emphasizing the significance of the direct methanol production process.

According to Le Chatelier's principle, an increase in reaction pressure and a decrease in reaction temperature results in a reaction shift toward the products, and more CH_3OH formation takes place than CO formation (equation 2). In addition, the high molar ratio of H_2/CO_2 reactants favours the CH_3OH formation by taking the two equilibria into account. As a result, low temperature and high pressure are more advantageous for CH_3OH production (Figures 2a, b and c).

3. Monometallic Catalysts

Extensive surface science experiments have been made to determine the Cu surface structural sensitivity by analysing the CO_2 activity of single crystals such as Cu (111), (100), and Cu (110), which also shows the structural dependent activity of Cu on CO_2 hydrogenation [19, 20]. Copper (Cu) has shown the unique ability to catalyse CO_2 into value-added chemicals (e.g., methanol, methane, carbon monoxide). CO_2-TPD (temperature-programmed desorption) studies established that Cu metal favour CO_2 chemisorption and is considered an active site for CO_2 hydrogenation than the other two phases (Cu_2O, CuO) [19]. Rasmussen et al., observed the first-order dissociative adsorption of CO_2 on clean Cu (100) at 740 Torr in the range of 475-550 K [21]. Further studies with TPD techniques proved that vicinal Cu (332), and

Cu (311) surfaces are more capable of CO_2 dissociation into CO over flat Cu surfaces [22]. Surface defect sites such as steps, edges, and kinks are essential for CO_2 activation. These experiments were carried out in ultrahigh vacuum (UHV) conditions, which differ significantly from the actual catalytic reactions of high pressure and temperature. Considering the thermodynamic and kinetic effects, such a change in pressure and temperature may lead to different reactivity patterns compared to UHV. A recent study by baren et. al demonstrated that CO_2 could dissociatively adsorb on Cu (100) and Cu (111) surfaces even at ambient pressure [17]. The extent of the dissociative adsorption of CO_2 is greater for Cu (100) than for Cu (111), and it results in the formation of surface oxygen, which prevents further CO_2 adsorption. For the same reason, industries mix the CO in the reactants to remove the surface oxygen.

The carbon-labelling experiments confirm that CO_2 is the source of the carbon in methanol [23, 24]. Ambient pressure x-ray photoelectron spectroscopy (APXPS) studies of CO_2 dissociation on Cu (997) metal surface at 340 K infer the formation of carbonate by CO_2 activation and increase in gas-phase CO_2 binding energy [25]. CO_2 dissociation results in the production of atomic oxygen, which reacts with CO_2 to generate CO_3 on the surface. This indicates that the adsorbed carbonate negatively charged by charge transfer from the Cu (997) surface results in the increase of work function. The present findings elucidate the significance of the carbonate intermediate in CO_2 chemistry on a stepped Cu surface. Overall, experiments on monometallic Cu surfaces show dissociative CO_2 adsorption, the importance of stepped Cu surfaces, and the formation of negatively charged carbonate species on Cu (997) surfaces.

4. Bimetallic Catalysts

Recent operando studies indicated that the Cu-Zn interface plays a crucial part in CO_2 hydrogenation that results in synergistic interactions between Cu and Zn. Various studies have shown that methanol production increases when Zn is near Cu. Introducing Zn atoms on Cu step sites has been shown to stabilize adsorption intermediates with lower energy barriers, as reported by Behrens et al., [26, 27]. This results in a considerably higher methanol production rate on the Cu-Zn surface compared to the unmodified Cu surface. As shown in Figure 3, 0.1 ML Zinc deposited on Cu thin film show some of the chemisorbed $CO_2^{\delta-}$ converted to carbonate (CO_3^{-2}) [28]. An addition of H_2O

molecules with CO_2 reactant, and the formation of methoxy, hydroxyl and formate species were observed, which provides the hydrogen source for the reaction. Control experiments on the Zn-deposited Cu (997) and its comparison with bare Cu (997) showed that zinc stabilizes the reaction products ($HCOO^-$ and CO_3^-) and creates additional adsorption sites, resulting in greater reaction product coverage [25, 28]. Thus activated CO_2 on the Cu metal surface needs extra support or promoter metal to stable the intermediate species and convert it to methanol.

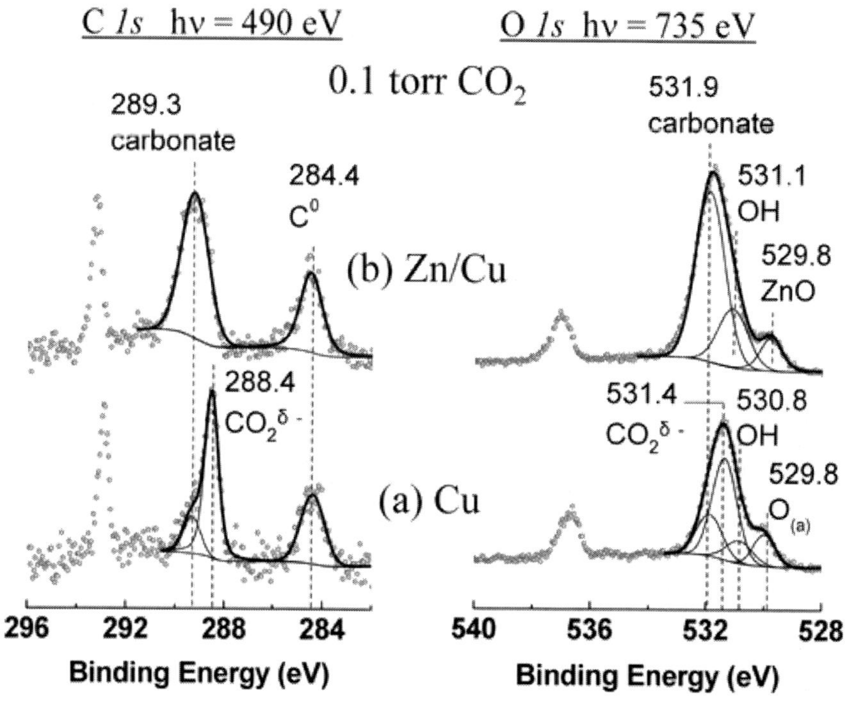

Figure 3. C 1s and O 1s XPS spectra of (a) pure Cu and (b) Zn/Cu (0.1 ML Zn) in the presence of 0.1 torr of CO_2 at room temperature. Three carbon species were identified on Cu: carbonate (CO_3^{2-}) at 289.3 eV, chemisorbed CO_2 ($CO_2^{\delta-}$) at 288.4 eV, and graphitic and/or hydrogenated C residues (C^0) at 284.4 eV. On the zinc-covered Cu surface carbonate is formed instead of $CO_2^{\delta-}$. In the oxygen region, the species identified on Cu were carbonate (CO_3^{2-}) at 531.9 eV, chemisorbed CO_2 ($CO_2^{\delta-}$) at 531.4 eV, surface hydroxyl (OH) at 530.8 eV, and chemisorbed oxygen at 529.8 eV. On Zn/Cu, the peak due to carbonate (CO_3^{2-}, 531.9 eV) dominates the spectrum. Adapted with permission from reference [28]. Copy right 2008 American Chemical Society.

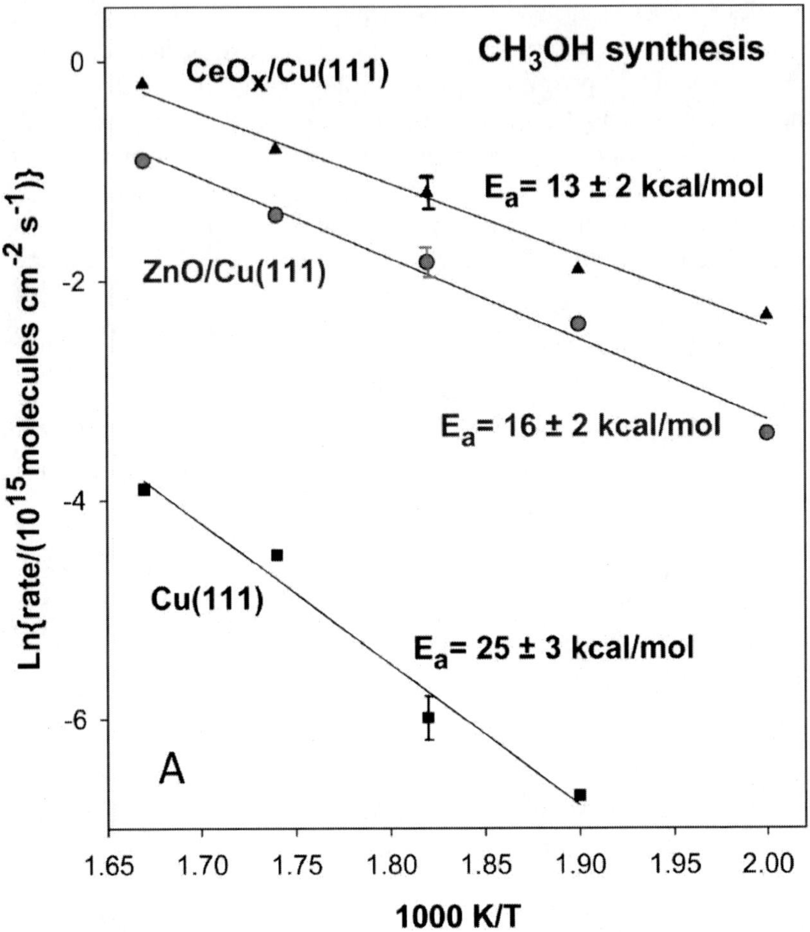

Figure 4. Arrhenius plots for the conversion of CO_2 to methanol on plain Cu (111) and Cu (100) and on metal surfaces covered ~20% by nanoparticles of ZnO. Reaction Conditions: $P\ H_2 = 4.5$ atm, $P\ CO_2 = 0.5$ atm. Adapted with permission from reference [35]. Copy right 2018 American Chemical Society.

Despite the prominence of the Cu-Zn interface in CH_3OH production, the nature of the catalyst's active sites remain elusive; two feasible active sites have been suggested in the literature. The first one is due to the synergetic interaction between Cu and ZnO at the interface, [29] and the second one is the Cu-Zn alloy formation [30]. Kattel et al., made progress toward the first possibility by comparing the ZnO/Cu (111) and ZnCu (111) model catalysts. A combination of experimental and theoretical data reveals that the ZnCu (111) experiences oxidation under given reaction conditions, resulting in the surface Zn being oxidized to ZnO. This leads to comparable activity to the ZnO/Cu (111) surface. Hence, the synergetic interaction at the interface should be responsible for methanol production [31]. As an example of second possible active site, e.g., Behrens et al., found that the Zn-atom-decorated Cu steps are the active sites, and its stability depends on the co-existence of well-defined surface species and bulk defects [26]. Additionally, Karelovic et al., have investigated the effect of Cu particle size on CO_2 hydrogenation, and they have emphasized the importance of surface steps, edges, and kinks in smaller nanoparticles [32-34]. The Arrhenius plot for the methanol production on ZnO/Cu(100) and ZnO/Cu(111) was investigated by Palamino et al., and shown in Figure 4. ZnO/Cu(100) is a superior catalyst for methanol synthesis in the temperature range of 500-600 K. On Cu(100), the activation energy for this reaction is 19 kcal/mol, whereas on ZnO/Cu(100), it drops to 11 kcal/mol. The rate at which methanol is produced is increased by 15-45 times on ZnO/Cu(100) compared to Cu(100) [35].

The strong metal-support interaction (SMSI) and adsorbate-induced oxidation cause a partial reduction of Zn to $Zn^{\delta+}$. XPS results corroborate the SMSI effect by revealing an inhomogeneous Zn distribution between bulk and surface, with a Zn-rich (Zn/Cu = 70/30) surface and a Cu-rich (Zn/Cu = 30/70) bulk [26]. Further, CO-TPD and in-situ FTIR studies revealed such Cu and ZnO interaction (SMSI) [36, 37]. DFT studies shows that HCO*, H_3CO*, and H_2CO* intermediates adsorption strength can be increased by the addition of Zn [26]. Thus, high activity with Cu/ZnO catalysts can be achieved through a combination of two factors: a) Steps on the Cu surface, which can be stabilized by bulk defects such as stacking faults or twin boundaries that terminate at the surface, and b) the presence of $Zn^{\delta+}$ at the defective Cu surface, which is the outcome of a dynamic SMSI effect.

Further, transmission electron microscopy (TEM) experiments on commercial Cu-ZnO-Al_2O_3 catalysts illustrated the SMSI-induced metastable ZnOx layer formation over Cu nanoparticles [26]. The active sites in the Cu-Zn nano ensemble have unique nanostructured features that allow for selective CO_2 conversion to methanol. It shows many Cu atoms in microcrystalline domains which are undercoordinated. This is especially for Cu atoms located near steps, edges, discontinuities, and planar boundaries. Several distinct types of partially reduced ZnOx overlayers are generated on Cu microdomains, and the SMSI continues to be an important role in active site generation. These findings highlight the relevance of regulating nanoscale characteristics of methanol synthesis catalysts.

The activation of CO_2 requires high pressure thereby leading to an increase in the required amount of energy, hence the ambient pressure route is anticipated for commercial production. In addition, low-pressure reaction conditions can make the operation safe and can be easily combined with the upstream feedstock-producing processes. However, thermodynamic, and kinetic considerations limit the use of low-pressure technologies to convert CO_2/H_2 into methanol. Many efforts have been made to develop a catalyst that can operate at ambient pressure; however, very few reports have demonstrated a high CH_3OH production rate at reaction pressure below 2 MPa. Generally, the rate of RWGS is higher than that of methanol synthesis anything under below 2 MPa leads to very poor selectivity towards methanol [34]. Some of the reports demonstrated nearly 100% methanol selectivity at/below 2 MPa using Cu-ZnO_x supported on SiO_2 catalysts [38]. Although these catalysts showed excellent methanol selectivity, poor CO_2 conversion (2 to 5%) results in a low methanol production yield.

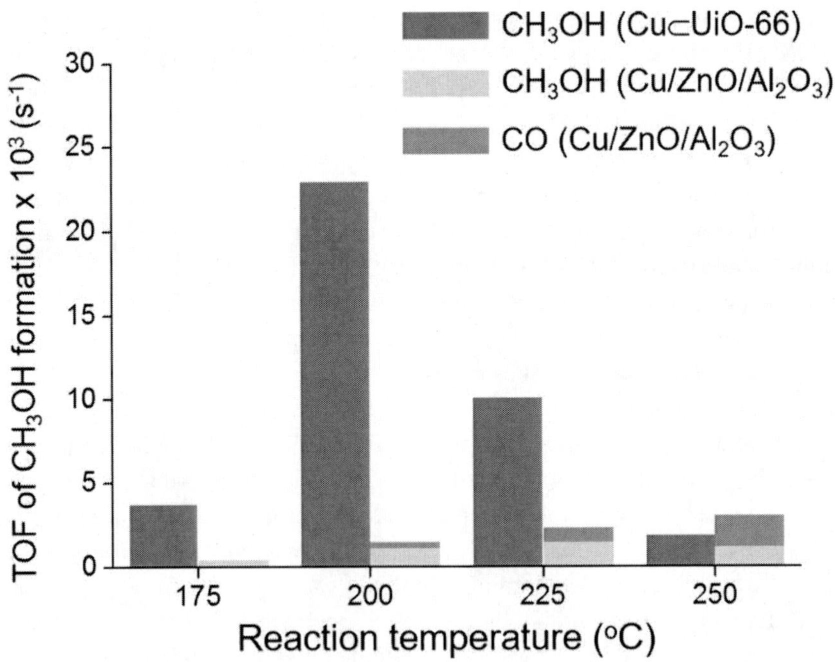

Figure 5. TOFs of product formation over Cu⊂UIO-66 catalyst and Cu/ZnO/A12O3 catalyst as various reaction temperatures. No CO is produced in the case of Cu⊂UiO-66 under all reaction temperatures. Adapted with permission from reference [40]. Copy right 2016 American Chemical Society.

Recent research has shown that the same Cu-ZnO materials with improved catalysts had a methanol selectivity ranging between 30% to 70% and CO_2 conversions reported to be less than 30% under typical reaction conditions (Pressure: <5 MPa, Temperature: 220-300 °C, H_2/CO_2: 3) [26, 31]. With further increase of reaction pressure (36 MPa) and reactant molar ratio (H_2/CO_2: 10), direct conversion of CO_2 to methanol increased to 95.3%, and 98.2% selectivity was obtained [39]. Copper-based metal-organic framework (MOF) (e.g., CuZn@UiO-bpy, Cu-UiO-66) showed 100% methanol selectivity and enhanced activity [40, 41]. However, the stability of these catalysts limits their wide usage (Figure. 5). Other Cu-based catalysts such as Cu_x-CeO_2, Ni-Ga_2O_3, Cu-Zr and Ni-Ga/SiO_2, have shown high efficiency in methanol synthesis at relatively low pressure [42-45]; however, the methanol yields are not good. Chen et al., synthesized high Cu dispersed particles on La-modified SBA-15 support by optimizing the interaction between Cu and LaOx. In comparison with Cu supported on SBA-15 catalyst, optimized $Cu_1La_{0.2}$/SBA-15 showed stable activity over 100 h with 81.5% methanol selectivity. In-situ DRIFTS results further shows the *HCOO and *OCH_3 species formation as essential intermediates over the Cu1$La_{0.2}$/SBA-15 catalyst surface under reaction conditions [46].

Recently, Coperet et al., synthesized the small and highly dispersed Cu nanoparticles alloyed with Ga on SiO_2-containing residual Ga III sites by surface organometallic chemistry approach. This method allows the synthesis of small distributed CuGax nanoparticles on silica support in conjunction with residual Ga III Lewis acidic sites. X-ray absorption spectroscopy reveals that gallium gets oxidized during reactions and copper remains the same as Cu^0. $CuGa_x$ stabilized the methoxy species on the surface while no formate was detected using ex-situ IR and solid-state NMR (nuclear magnetic resonance). These $CuGa_x$ catalysts exhibit improved CO_2 hydrogenation activity over Cu-Zr/SiO_2 and Cu-Ti/SiO_2 catalysts, with overall CH_3OH (CH_3OH + DME) selectivity of up to ca. 90% with 3% conversion [47, 48].

While Cu-based catalysts continue to hold great promise for the methanol synthesis, further improvement in the performance is needed to overcome the technological issues. So far, many studies from around the world have devoted significant time and effort to unveil the CO_2 chemistry and improving catalytic performance. In terms of reaction thermodynamics, reaction favors at low temperature and high pressures; however, breaking the O = C = O bonds in CO_2 molecules at low temperatures is kinetically unfavorable. While high temperatures favor RWGS, high pressures limit process efficiency. It is difficult to achieve both high CO_2 conversion and high methanol selectivity with the present $Cu/ZnO/Al_2O_3$ catalysts due to the sintering of Cu particles under reaction conditions. Hence, strong metal support interaction and high thermal stability are expected to improve the catalysts life and can be achieved by adding the promoters to make bimetallic/multi-metallic catalysts, synthesizing the confined catalysts, core-shell structures, etc. In addition to adopting the suitable conventional synthesis methods, developing the new methods are promising for improving the methanol selectivity. Despite of catalyst active sites and reaction mechanisms been investigated in the literature, a detailed structural understanding of the catalysts remains elusive and modern analytical techniques such as APXPS, and EXAF needs to be deployed to unravel and establish the structure and activity relationship. Therefore, a complete approach incorporating fundamental knowledge of catalysis with sophisticated catalyst synthesis and process optimization is required to improve the methanol selectivity.

5. Reaction Pathways and Intermediates

As discussed, the CO_2 hydrogenation to methanol can proceed either direct or indirect way. The indirect way has two mechanistic paths based on the key reaction intermediates i) the formate pathway ($HCOO^-$) and ii) the hydrocarboxyl pathway (COOH). Many studies have proposed a formate-mediated mechanism for methanol synthesis as it has a lower barrier than the hydrocarboxyl formation [31, 49, 50]. Moreover, the formate intermediates were determined to be the maximum available surface species under the CO_2 hydrogenation [51]. According to DFT calculations and microkinetic modeling studies on the Cu surface found the formate pathway for methanol synthesis proceeds with HCOO*, HCOOH*, CH_3O_2*, CH_2O*, and CH_3O* species [49]. The presence of low-coordinated Cu atoms in nanoparticles aids in the stabilization of the key intermediates (HCOO and H_2COO), lowering the rate-limiting hydrogenation process. Some studies show the water-promoted hydrocarboxyl-mediated methanol synthesis with COOH → HCOOH → HCO → HCOH → CH_2OH → CH_3OH pathway is more kinetically favorable than the formate pathway on Cu (111) [52]. Hydrocarboxyl is also believed to be a precursor of CO in the RWGS process. A second proposed method for indirect CO_2 hydrogenation to methanol begins with the production of CO (RWGS), followed by hydrogenation to form methanol. However, the indirect pathway is not widely accepted as direct pathway over Cu-based catalyst, as CO_2 hydrogenation to methanol was found to occur substantially faster than CO hydrogenation to methanol over the industrial Cu/ZnO catalyst. Furthermore, isotopic labeling studies have already established that methanol is mostly produced through the hydrogenation of CO_2 rather than CO [53, 54].

6. Catalyst Preparation

Many studies have been reported that metal-oxide interfaces play an active part in the catalysis, as each component could offer a specific property that is essential for catalytic activity [55-57]. These studies showed us the importance of the metal-oxide interface and how to regulate for crucial processes such as CO_2 hydrogenation, CO oxidation, CH_3OH oxidation, etc. and the increase of metal dispersion/active surface area wich is achieved by varying the synthesis method, adding promoters, high surface area support, and changing the metal precursor. In CO_2 hydrogenation, single metal copper supported on metal oxide catalysts are limited due to low stability and poor methanol selectivity. An addition of promoter MO_x (M = Ce, Mn, Ga, Zr, Si, Ti, and Mg) to the Cu-based catalysts showed a positive impact on the catalytic performance due to the high dispersion of Cu-MOx active species [42, 58-60]. Li et al., produced mesoporous CuO-ZnO-ZrO_2 material using a surfactant-assisted co-precipitation approach as an alternative to the traditional method [61]. Small Cu particle dispersion on mesoporous structures was discovered to be responsible for keeping the large interfacial area of the Cu-ZnO_x and/or Cu-ZrO_x interfaces, which in turn led to a higher turnover frequency (TOF) for methanol production. Thus, the goal of catalysts synthesis is to increase the spatial distribution of the metal nanoparticles on the support. The high surface area of the support allows metal particles to be dispersed across its surface, granting greater metal utilization and, as a result, a greater number of active sites per mass of catalyst. Nevertheless, difficulties with metal aggregation and sintering during catalyst synthesis typically result in less-than-ideal catalytic systems. Particularly when expensive precious group metals are the active sites, the loss of dispersion is undesirable. Although various methods have been adopted to synthesize the Cu-Zn-based catalysts, here we have discussed the commonly used methods in the following sections.

6.1. Impregnation Method

Incipient wetness impregnation (IWI), often known as dry impregnation, is the most widely employed synthesis method (DI). Here, support is impregnated with a metal precursor that has been dissolved in water, which should equal to the pore volume of the support. As a result, no metal is wasted during the synthesis process. Hence this is an attractive method for preparing noble metal-supported catalysts. This is a very simple method as no filtration is required and a known quantity of metal can be loaded on the support. However, it is difficult to tune the metal dispersion or promote bimetallic interaction with this method, resulting in larger particle sizes and a weak bimetallic character. In addition, the pores or channels in supports hinder the complete wetting of the surface with the impregnating solution. As a result, the entire support surface area is not utilized, and agglomeration can occur in support channels.

6.2. Co-Precipitation Method

This method is widely used in the synthesis of bimetallic and multi-metallic supported on metal oxide catalysts or mixed oxide catalysts. It all starts with the simultaneous precipitation of a metal precursor in a suitable solvent by adjusting its pH, which gives an equally distributed catalyst precursors with a known composition. The high dispersion of catalyst components can be obtained by the coprecipitation process, which is challenging to attain through other methods.

Conclusion

This chapter summarized the developments in CO_2 chemistry in terms of Cu-based bimetallic catalysts for CO_2 hydrogenation to methanol. Advancement in operando techniques has led to a better understanding of heterogeneous catalysts for CO_2 hydrogenation, particularly the Cu-based system. The surface science investigations on Cu surfaces revealed that metallic copper has an active site for CO_2 chemisorption, which aided in developing a novel and more effective catalyst. Monometallic Cu surfaces show dissociative CO_2 adsorption, the importance of stepped Cu surfaces, and the formation of negatively charged carbonate species on Cu (997) surfaces. Further investigations into the utilization of monometallic Cu catalysts are limited due to the poor stability of active intermediate species for methanol selectivity. Adding a promoter metal to the Cu, especially Zn catalysts had an interfacial interaction, and creates an active site, which enhances the methoxy species formation and positive impact on the CO_2 hydrogenation to methanol. Significant progress has been made in the development of supported metal catalysts for CO_2 hydrogenation, including Cu/ZnO/Al_2O_3. This is regarded as providing active sites for spillover hydrogen or promote the dispersion of other components. A combination of experimental and theoretical studies reveals that the synergetic interaction between Cu and ZnO creates new active sites and promotes the metal dispersion responsible for the methanol production. Hence, each component offers a characteristic property that is essential for effective catalysis in intricate reaction paths. Moreover, HCOO intermediates were determined to be the most available surface species during the CO_2 hydrogenation reaction. Currently, various approaches like increasing the Cu dispersion, adding more promoters, and adopting new synthesis methods are the main ways to improve the performance of CO_2 hydrogenation to methanol.

References

[1] Daza, Y. A., Kuhn, J. N. CO_2 conversion by reverse water gas shift catalysis: comparison of catalysts, mechanisms, and their consequences for CO_2 conversion to liquid fuels, *RSC Adv.* 2016, *6,* 49675-49691.

[2] Olah, G. A. Towards Oil Independence Through RenewableMethanol Chemistry, *Angew. Chem. Int. Ed.* 2013, *52*, 104-107.

[3] Torrente-Murciano, L., Mattia, D., Jones, M. D., Plucinski, P. K. Formation of hydrocarbons via CO_2 hydrogenation-A thermodynamic study, *J. CO_2 Util.* 2014, *6,* 34-39.

[4] Niu, J., Liu, H., Jin, Y., Fan, B., Qi, W., Ran, J. Comprehensive review of Cu-based CO_2 hydrogenation to CH_3OH: Insights from experimental work and theoretical analysis *Int. J. Hydrog. Energy* 2022, *47,* 9183-9200.

[5] María, R. D., Diaz, Ismael., Rodriguez, M., Saiz, A. Industrial Methanol from Syngas: Kinetic Study and Process Simulation, *Int. J. Chem. React.* 2013, *11,* 469-477.

[6] Yan, B., Zhao, B., Kattel, S., Wu, Q., Yao, S., Su, D., Chen, J. G. Tuning CO_2 hydrogenation selectivity via metal-oxide interfacial sites, *J. Catal.* 2019, *374,* 60-71.

[7] Jiang, X., Nie, X., Guo, X., Song, C., Chen, J. G. Recent Advances in Carbon Dioxide Hydrogenation to Methanol via Heterogeneous Catalysis, *Chem. Rev.* 2020, *120*, 7984-8034.

[8] Reddy, K. P., Dama, S., Mhamane, N. B., Ghosalya, M. K., Raja. T., Satyanarayana, C. V., Gopinath, C. S. Molybdenum carbide catalyst for the reduction of CO_2 to CO: surface science aspects by NAPPES and catalysis studies, *Dalton Trans.* 2019, *48*, 12199-12209.

[9] Lee, D. E., Naresh, M., Reddy, K. P., Moses, B. M., Jo, W-K., Tonda, S. Bifunctional S-scheme hybrid heterojunction comprising CdS nanorods and $BiOIO_3$ nanosheets for efficient solar-induced antibiotic degradation and highly-selective CO_2 reduction, *J. Mater. Sci Technol. 2023*, *161*, 74-87.

[10] Patra, K. K., Prak, S., Song, H., Kim, B., Kim, W., Oh, J. Operando Spectroscopic Investigation of a Boron-Doped CuO Catalyst and Its Role in Selective Electrochemical C-C Coupling, *ACS Appl. Energy Mater.* 2020, *3*, 11343-11349.

[11] Tudu, B., Nalajala, N., Reddy, K. P., Saikia, P., Gopinath, C. S. Rationally Designed, Efficient, and Earth-Abundant Ni-Fe Cocatalysts for Solar Hydrogen Generation, *ACS Sustainable Chem. Eng.* 2021, *9,* 13915-13925.

[12] Tudu, B., Nalajala, N., Reddy, K. P., Saikia, P., Gopinath, C. S. Electronic Integration and Thin Film Aspects of Au-Pd/rGO/TiO_2 for Improved Solar Hydrogen Generation, *ACS Appl. Mater. Interfaces* 2019, *11,* 32869-32878.

[13] Zhan, W., Wang, J., Wang, H., Zhang, J., Liu, X., Zhang, P., Chi, M., Guo, Y., Guo, Y., Lu, G., Sun, S., Dai, S., Zhu, H. Crystal Structural Effect of AuCu Alloy Nanoparticles on Catalytic CO Oxidation, *J. Am. Chem. Soc.* 2017, *139,* 8846-8854.

[14] Wang, L. X., Guan, E., Wang, Z., Wang, L., Gong, Z., Cui, Y., Yang, Z., Wang, C., Zhang, J., Meng, X., Hu, P., Gong, X-Q., Gates, B. C., Xiao, F-S. Dispersed Nickel Boosts Catalysis by Copper in CO_2 Hydrogenation, *ACS Catal.* 2020, *10*, 9261-9270.

[15] Li, G., Zhang, W., Luo, N., Xue, Z., Hu, Q., Zeng, W., Xu, J. Bimetallic Nanocrystals: Structure, Controllable Synthesis and Applications in Catalysis, Energy and Sensing, *Nanomaterials* 2021, *11*, 1926.

[16] Wolfbeisser, A., Kovacs, G, Kozlov, S. M., Fottinger, K., Bernardi, J., Klotzer, B., Neyman, K. M., Rupprechter, G. Surface composition changes of CuNi-ZrO_2 during methane decomposition: An *operando* NAP-XPS and density functional study, *Catal. Today* 2017, *283*, 134-143.

[17] Eren, B., Weatherup, R. S., Liakakos, N., Somorjai, G. A., Salmeron, M. Dissociative Carbon Dioxide Adsorption and Morphological Changes on Cu(100) and Cu(111) at Ambient Pressures, *J. Am. Chem. Soc.* 2016, *138*, 8207-8211.

[18] Roy, K., Gopinath, C. S. UV Photoelectron Spectroscopy at Near Ambient Pressures: Mapping Valence Band Electronic Structure Changes from Cu to CuO, *Anal. Chem.* 2014, *86,* 3683-3687.

[19] Rasmussen, P. B., Taylor, P. A., Chorkendorff, I, The interaction of carbon dioxide with Cu(100), *Surf. Sci* 1992. *269-270,* 352-359.

[20] Yoshihara, J., Campbell, C. T., Methanol Synthesis and Reverse Water-Gas Shift Kinetics over Cu(110) Model Catalysts: Structural Sensitivity, *J. Catal.* 1996, *161,* 776-782.

[21] Rasmussen, P. B., Holmblad, P. M., Askgaard, T., Ovesen, C. V., Stoltze, Norskov, J. K., Chorkendorff, I, Methanol synthesis on Cu(100) from a binary gas mixture of CO_2 and H_2, *CataL. Lett.* 1994, *26,* 373-381.

[22] Bonicke, I. A., Kirstein, W., Thieme, F., A study on CO_2 dissociation on a stepped (332) copper surface, *Surf. Sci.* 1994, *307-309,* 177-181.

[23] Chinchen, G. C., Denny, P. J., Parker, P. J., Spencer, M. S., Whan, D. A. Mechanism of methanol synthesis from $CO_2/CO/H_2$ mixtures over copper/zinc oxide/alumina catalysts: use of ^{14}C-labelled reactants, *Appl. Catal.* 1987, *30,* 333-338.

[24] Chinchen, G. C., Denny, P. J., Jennings, J. R., Spencer, M. S., Waugh, K. C. Synthesis of Methanol: Part 1. Catalysts and Kinetics, *Appl. Catal.* 1988, *36,* 1-65.

[25] Koitaya, T., Yamamoto, S., Shiozawa, Y., Yoshikura, Y., Hasegawa, M., Tang, J., Takeuchi, K., Mukai, Koji., Yoshimoto, S., Matsuda, I., Yoshinobu, J., CO_2 Activation and Reaction on Zn-Deposited Cu Surfaces Studied by Ambient-Pressure X-ray Photoelectron Spectroscopy, *ACS Catal.* 2019, *9*, 4539-4550.

[26] Behrens, M., Studt, F., Kasatkin, S., Kuhl, S., Havecker, M., Abild-Pedersen, F., Zander, S., Girgsdies, F., Kurr, P., Kniep, B-L., Tovar, M., Fischer, R. W., Norskove, J. K., Schlogl, R. The Active Site of Methanol Synthesis over $Cu/ZnO/Al_2O_3$ Industrial Catalysts, *Science* 2012, *336*, 893-897.

[27] Studt, F., Behrens, M., Kunkes, E. L., Thomas, N., Zander, S., Tarasov, A., Schumann, J., Frei, E., Varley, J. B., Abid-Pedersen, F., Norskov, J. K., Schlogl, R. The Mechanism of CO and CO_2 Hydrogenation to Methanol over Cu-Based Catalysts, *ChemCatChem* 2015, *7,* 1105-1111.

[28] Deng, X., Verdaguer, A., Herranz, T., Weis, C., Bluhm, H., Salmeron, M. Surface Chemistry of Cu in the Presence of CO_2 and H_2O, *Langmuir* 2008, *24*, 9474-9478.

[29] Nakamura, J., Uchijima, T., Kanai, Y., Fujitani, T. The role of ZnO in Cu/ZnO methanol synthesis catalysts, *Catal. Today* 1996, *28,* 223-230.

[30] Fujitani, T., Nakamura, J., The chemical modification seen in the Cu/ZnO methanol synthesis catalysts, *Appl. Catal. A* 2000, *191,* 111-129.

[31] Kattel, S., Ramirez, P. J., Chen, J. G., Rodriguez, J. A., Liu, P. Active sites for CO_2 hydrogenation to methanol on Cu/ZnO catalysts, *Science* 2017, *355*, 1296-1299.

[32] Karelovic, A., Galdames, G., Medina, J. C., Yevenes, C., Baraa, Yanitza, Jimenez, R. Mechanism and structure sensitivity of methanol synthesis from CO_2 over SiO_2-supported Cu nanoparticles, *J. Catal.* 2019, *369*, 415-426.

[33] Karelovic, A., Ruiz, P. The role of copper particle size in low pressure methanol synthesis *via* CO_2 hydrogenation over Cu/ZnO catalysts, *Catal. Sci. Technol.* 2015, *5*, 869-881.

[34] Yang, Y., Evans, J., Rodriguez, J. A., White, M. G., Liu, P. Fundamental studies of methanol synthesis from CO_2 hydrogenation on Cu (111), Cu clusters, and Cu/ZnO (0001), *Phys. Chem. Chem. Phys.,* 2010, *12,* 9909-9917.

[35] Palomino, R. M., Ramirez, P. J., Liu, Z., Hamlyn, R., Waluyo, I., Mahapatra, M., Orozco, I., Hunt, A., Simonovis, J. P., Senanayake, S. D., Rodriguez, J. A. Hydrogenation of CO_2 on ZnO/Cu(100) and ZnO/Cu(111) Catalysts: Role of Copper Structure and Metal-Oxide Interface in Methanol Synthesis, *J. Phys. Chem. B.* 2018, *122*, 794-800.

[36] Naumann d'Alnoncourt, R., Xia, X., Strunk, J., Loffler, E., Hinrichsen, O., Muhler, M. The influence of strongly reducing conditions on strong metal-support

interactions in Cu/ZnO catalysts used for methanol synthesis, *Phys. Chem. Chem. Phys.* 2006, *8,* 1525-1538.

[37] Topsøe, N. Y., Topsøe, H. On the nature of surface structural changes in Cu/ZnO methanol synthesis catalysts, *Top Catal.* 1999, *8*, 267-270.

[38] Toyir, J., Piscina, P. R. D. L., Fierro, J. L. G., Homs, N. Highly effective conversion of CO_2 to methanol over supported and promoted copper-based catalysts: influence of support and promoter, *Appl. Catal. B*, 2001, *29*, 207-215.

[39] Bansode, A., Urakawa, A. Towards full one-pass conversion of carbon dioxide to methanol and methanol-derived products, *J. Catal.* 2014, *309,* 66-70.

[40] Rungtaweevoranit, B., Baek, J., Araujo, J. R., Archanjo, B. S., Choi, K. M., Yagi, O. M., Somorjai, G. A. Copper Nanocrystals Encapsulated in Zr-based Metal-Organic Frameworks for Highly Selective CO_2 Hydrogenation to Methanol, *Nano Lett.* 2016, *16*, 7645-7649.

[41] An, B., Zhang, J., Cheng, K., Ji, P., Wang, C., Lin, W. Confinement of Ultrasmall Cu/ZnO_x Nanoparticles in Metal-Organic Frameworks for Selective Methanol Synthesis from Catalytic Hydrogenation of CO_2, *J. Am. Chem. Soc.* 2017, *139*, 3834-3840.

[42] Wang, W., Qu, Z., Song, L., Fu, Q., CO_2 hydrogenation to methanol over Cu/CeO_2 and Cu/ZrO_2 catalysts: Tuning methanol selectivity via metal-support interaction, *J. Energy Chem.* 2020, *40*, 22-30.

[43] Choi, H., Oh. S., Tran, S. B. T., Park, J. Y., Size-controlled model Ni catalysts on Ga_2O_3 for CO_2 hydrogenation to methanol, *J. Catal.* 2019, *376*, 68-76.

[44] Wu, C., Lin, L., Zhang, J., Zhang, F. Zhou, T., Rui, N., Yao, S., Deng, Y., Yang, F., Xu, W., Luo, J., Zho, Y., Yan, B., Wen, X. D., Rodriguez, J. A., Ma, D., Inverse ZrO_2/Cu as a highly efficient methanol synthesis catalyst from CO_2 hydrogenation, *Nat. Commun.* 2020, *11,* 5767.

[45] Gallo, A., Snider, J. L., Sokaras, D., Nordlund, D., Kroll, T., Ogasawara, H., Kovarik, L., Duyar, M. S., Jaramillo, T. F., Ni_5Ga_3 catalysts for CO_2 reduction to methanol: Exploring the role of Ga surface oxidation/reduction on catalytic activity, *Appl. Catal. B* 2020, *267*, 118369.

[46] Chen, K., Fang, H., Liu, X., Zheng, J., Zhou, S., Duan, X., Zhuang, Y., Tsang, S. C. E., Yuan, Y. CO_2 hydrogenation to methanol over Cu catalysts supported on La-modified SBA-15: The crucial role of Cu-LaO_x interfaces, *Appl. Catal. B* 2019, *251*, 119-129.

[47] Lam, E., Larmier, K., Wolf, P., Tada, S., Safanova, O. S., Coperet, C. Isolated Zr Surface Sites on Silica Promote Hydrogenation of CO_2 to CH_3OH in Supported Cu Catalysts, *J. Am. Chem. Soc.* 2018, *140*, 10530-10535.

[48] Lam, E., Noh, G., Chan, K. W., Larmier, K., Lebedev, D., Searles, K., Wolf. P. Safonova, O. V., Coperet. C. Enhanced CH_3OH selectivity in CO_2 hydrogenation using Cu-based catalysts generated *via* SOMC from Ga^{III} single-sites, *Chem. Sci.* 2020, *11*, 7593-7598.

[49] Grabow, L. C., Mavrikakis, M. Mechanism of Methanol Synthesis on Cu through CO_2 and CO Hydrogenation, *ACS Catal.* 2011, *1,* 365-384.

[50] Yatsu, T., Nishimura, H., Fujitani, T., Nakamura, J. Synthesis and Decomposition of Formate on a Cu/SiO_2 Catalyst: Comparison to Cu (111), *J. Catal.* 2000, *191,* 423-429.

[51] Fujitani, T., Nakamura, I., Uchijima, T., Nakamura, The kinetics and mechanism of methanol synthesis by hydrogenation of CO_2 over a Zn-deposited Cu (111) surface, *J. Surf. Sci.* 1997, *383*, 285-298.

[52] Zhao, Y. F., Yang, Y., Mims, C., Peden, C. H. F., Li, J., Mei, D. Insight into methanol synthesis from CO_2 hydrogenation on Cu(111): Complex reaction network and the effects of H_2O, *J. Catal.* 2011, *281*, 199-211.

[53] Hartadi, Y., Widmann, D., Behm, R. J., Methanol synthesis *via* CO_2 hydrogenation over a Au/ZnO catalyst: an isotope labelling study on the role of CO in the reaction process, *Phys. Chem. Chem. Phys.* 2016, *18*, 10781-10791.

[54] Kunkes, E. L., Sturd, F., Abild-Pedersen, F., Schlogl. R., Behrens, M. Hydrogenation of CO_2 to methanol and CO on Cu/ZnO/Al_2O_3: Is there a common intermediate or not? *J. Catal.* 2015, *328*, 43-48.

[55] Reddy, K. P., Choi, H., Kim, D., Ryoo, R., Park, J. Y. Cu oxide deposited on shape-controlled ceria nanocrystals for CO oxidation: influence of interface-driven oxidation states on catalytic activity, *Catal. Sci. Technol.* 2021, *11*, 6134-6142.

[56] Reddy, K. P., Choi, H., Kim, D., Choi, M., Ryoo, R., Park, J. Y. The facet effect of ceria nanoparticles on platinum dispersion and catalytic activity of methanol partial oxidation, *Chem. Commun.* 2021, *57*, 7382-7385.

[57] Zhang, Y., Zhang, Z., Yang, X., Wang, R., Duan, H., Shen, Z., Li, L., Su, Y., Yang, R., Zhang, Y., Su, X., Huang, Y., Zhang, T. Tuning selectivity of CO_2 hydrogenation by modulating the strong metal-support interaction over Ir/TiO_2 catalysts, *Green Chem.* 2020, *22*, 6855-6861.

[58] Wang, S., Yang, J., Wang, S., Xiao, F., Zhao, N., Effect of Preparation Method on the Performance of Cu, *ChemCatChem* 2022, *14*, e202200957.

[59] Medina, J. C., Figueroa, M., Manrique, R., Pereira, J. R., Srinivasan, P. D., Bravo-Suarez, J. J., Medrano, V. G. B., Jimenez, R., Karelovic, A. Catalytic consequences of Ga promotion on Cu for CO_2 hydrogenation to methanol, *Catal. Sci. Technol.* 2017, *7*, 3375-3387.

[60] Chang, K., Wang, T., Chen, J. G. Hydrogenation of CO_2 to methanol over CuCeTiO_x catalysts, *Appl. Catal. B* 2017, *206*, 704-711.

[61] Li, L., Mao, D., Yu, J., Guo, X. Highly selective hydrogenation of CO_2 to methanol over CuO-ZnO-ZrO_2 catalysts prepared by a surfactant-assisted co-precipitation method, *J. Power Sources* 2015, *279,* 394-404.

Chapter 3

Bimetallic Nanoparticles for Electrocatalytic Oxygen Reduction and Urea Oxidation Reactions

Rashmi Chetry*
Shaheen P. Bhuyan*
and Pankaj Bharali†
Department of Chemical Sciences, Tezpur University, Napaam, Assam, India

Abstract

The oxygen reduction reaction (ORR) and the urea oxidation reaction (UOR) are the most vital reactions in the case of industrial and commercial applications of non-conventional energy conversion systems, such as the fuel cells (FCs). However, due to their slow kinetics, they need to be assisted by electrocatalysts (ECs) that are practically affordable, as opposed to precious ECs. To address this issue, various measures and approaches have been under exploration in recent times, one of which involves electrocatalysts (ECs) based on bimetallic nanoparticles. In this chapter, various catalysts based on bimetallic compositions, including less precious electrocatalysts (ECs) applicable for fuel cells (FCs), and their constructive synergistic roles, resulting from the combination of various intrinsic characteristics such as crystal planes, dopants, morphology, etc., in facilitating and improving factors responsible for electrochemical activities, and hence enhancing the overall electrocatalytic activity towards both ORR and UOR, have been discussed.

---

* Equally contributed to the chapter.
† Corresponding Author's Email: pankajb@tezu.ernet.in.

In: Bimetals
Editors: Pranjal Saikia, Jayanta Kumar Sarmah and Parikshit GogoiName
ISBN: 979-8-89113-496-6

Keywords: oxygen reduction reaction, urea oxidation reaction, fuel cells, electrocatalysts

1. Introduction

Emissions from gasoline and diesel engine vehicles comprise of greenhouse gases (GHGs) and air pollutants. These have an adverse impact on the air quality, human health, and the environment in total. Also, the dramatic increase in the rate of consumption of fossil fuels, is leading to severe energy crisis. Hence substantial emphasis on the development of innovative and clean energy conversion systems with higher efficiency than the conventional ones is the need of the hour. Over the recent decades among the various non-conventional energy conversion systems, fuel cells (FCs) have been considered to be ideally complementing the idea of high efficiency and sustainable development [1-4].

FCs are electrochemical cells that convert reactants continuously supplied from an external source into electrical energy. They are fundamentally open thermodynamic systems [5, 6]. The properties such as high efficiency (30-90% more efficiency than internal combustion engines), simplicity, low to no emission of noise are what makes it an eye catchy innovation. It is a promising alternative for rural areas where there is no such public grid and proper electrical wiring to provide power supply and also it reduces a huge amount of cost of wiring, etc. [6].

Table 1. Comparison of fuel cell with other power generating systems [5]

	Diesel Engine	Turbine generator	Photovoltaic	Wind turbine	Fuel cells
Capacity range	500 kW–50 MW	500 kW–5 MW	1 kW–1 MW	10 kW–1 MW	200 kW–2 MW
Efficiency	35%	29–42%	6–19%	25%	40–85%
Capital cost ($/kW)	200-350	450–870	6600	1000	1500-3000
O & M cost ($/kW)	0.005-0.015	0.005-0.0065	0.001–0.004	0.01	0.0019-0.0153

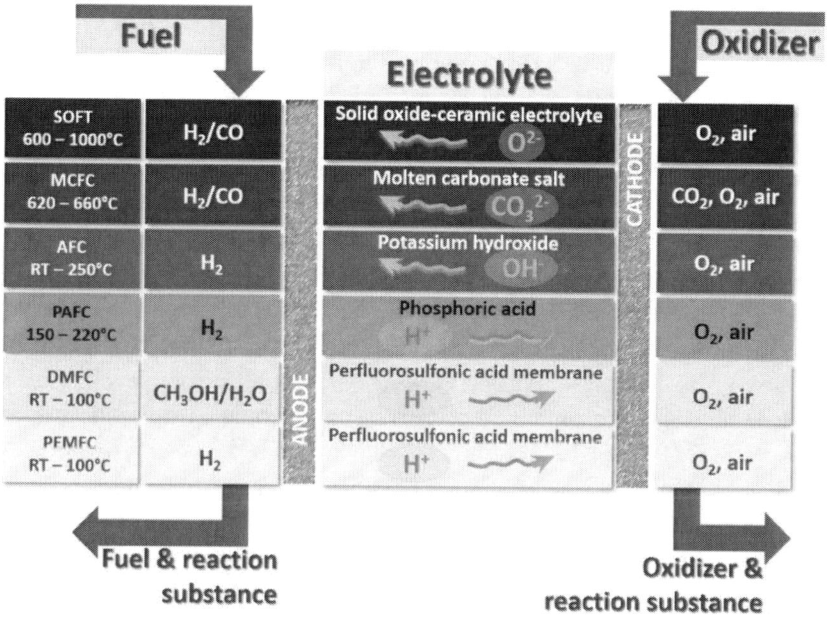

Figure 1. Overview of the various types of FCs with their operation temperatures. Adapted with permission from Reference [8], Copyright (2021), Multidisciplinary Digital Publishing Institute.

From Table 1 we can understand how FCs are much superior to the other power generating systems. Presently, FCs are employed widely in various applications involving both in small and large scale such as combined heat and power (CHP) systems, portable computers, military communication equipment, mobile computers, etc. Thanks to Sir William Grove, the inventor of the first FC which has now lead to today's advanced clean and green technologies enabling us to attain the goal of sustainable development [7].

There are various types of FCs available today and are categorized on the basis of the fuel and the electrolyte used and they are: i) alkaline fuel cell (AFC) ii) phosphoric acid fuel cell (PAFC) iii) solid oxide fuel cell (SOFC) iv) molten carbonate fuel cell (MCFC) v) proton exchange membrane fuel cell (PEMFC) vi) direct methanol fuel cell (DMFC) vii) direct urea fuel cell (DUFC) [8].

Figure 1 shows various types of FCs which operate on different temperatures. Additionally, direct urea fuel cells (DUFCs) are considered to be one of the most environmentally benign power generation as well as waste water treatment systems [9]. DUFCs are one of the alternatives to water splitting and hydrogen-oriented FCs [10]. However, they have been less studied and reviewed so far, as compared to the other types of FCs.

Various anodic and cathodic reactions are involved in different FCs which are fundamentally responsible for the generation of electricity. Some of the important reactions include oxygen reduction reaction (ORR) [11], oxygen evolution reaction (OER) [12], hydrogen evolution reaction (HER) [13], urea oxidation reaction (UOR) [14], formic acid oxidation reaction (FAOR) [15], methanol oxidation reaction (MOR) [16], etc. Reactions such as HER, FAOR, and UOR are also known to be potential sources for the production of H_2. H_2 has emerged to be one of the most popular alternatives to fossil fuel especially because of its no emission property. It also has an efficiency level 2-3 times higher than that of gasoline, hence owns its potential in domestic use and vehicles [17, 18].

Fast kinetics of the reactions involved in a FC ensures its better performance since with faster kinetics, availability of the required number of electrons increases, thereby making the process of generation of electricity more feasible. However, these reactions tend not to stand up to the mark due to their sluggish kinetics. Thus, assisting these electrochemical reactions with suitable catalysts in order to make them feasible for application in FCs is one of the basic requirements of a FC. In this chapter we will be emphasizing mainly upon the kinetics involved, types of electrocatalyst used and applications of ORR and UOR in FCs.

ORR is a cathodic reaction where the electrons generated at the anode are transferred through an external circuit to the cathode and the oxygen gets reduced to water or hydroxyl ions depending upon the electrolyte used i.e., acidic or alkaline, respectively [19]. The reaction proceeds via either of the two pathways: one pathway involves partial reduction of O_2 also known as "2 e^- pathway" and another involves the "4 e^- pathway" where O_2 is fully reduced. The former pathway involves the formation of hydrogen peroxide as an intermediate which hinders the kinetics, making it less efficient hence the "4 e^- pathway" is the more favoured pathway. In order to make the reaction suitable for a FC, the benchmark EC so far used is Pt/C. But commercializing FCs with the benchmark Pt/C catalyst takes up almost 36-50% of the FC's costing which is not practical, hence in the recent times researchers have been focused on developing suitable cost-effective and highly active ECs for their application in commercializing FCs [20].

On the other hand, UOR is an anodic reaction where oxidation of urea is preferentially done in alkaline electrolyte which involves a "6 e^- pathway." UOR is an underlying reaction which facilitates urea-based energy conversion systems including urea electrolysis for hydrogen production and in DUFCs for energy production. Another reason for UOR's popularity is because it also resolves the waste water disposal problem by the treatment of urea, as urea is considered to be a pollutant for both water and soil contamination. However, a similar case occurs here like in the case of ORR i.e., the benchmark EC reported for UOR also involves precious-inert-metal catalysts like RuO_2, Pt or Pd based which makes its commercialization impractical due to its cost. Thus, designing and fabrication of ECs which breaks the barrier of commercialization of FCs due to its high costing is the need of the hour. Through studies it has been found that certain complementing bimetallic ECs show better electrochemical activities, which is due to the synergistic effect between the two metals [21]. The term synergistic is related to the interaction and assistance of two or more substances to produce a combined effect which is greater than the sum of the individuals. In case of synergistic effect between two metals in a catalyst, the combination of the components can be like that of crystal planes, morphologies, impurities in the form of dopants, structures, support, etc. to obtain the merits of both the metals which in turn makes up for the disadvantage of a single compound, thereby enhancing the overall catalytic activity of the system [22].

Thus, in this chapter we will be discussing the recent advancements in synthesis of various less precious bimetallic EC with high electrochemical activity for its application in ORR and UOR.

2. Oxygen Reduction Reaction

As discussed earlier, the energy crisis has detrimental impacts on the environment, emphasizing the need for alternative sources. PEMFCs and Metal-air batteries (MABs) with high efficiency and environmentally benign characteristics have attracted much attention owing to their irreplaceable roles in the construction of the future sustainable energy system [23-24]. ORR is regarded as one of the crucial steps in many energy storage devices, including FCs and MABs [23-26]. The direct or full reduction pathway is more effective that follows $4e^-$ pathway and results in the formation of H_2O (Figure 2a). The other process can occur via an indirect or partial reduction pathway, using a $2e^-$ route to produce adsorbed H_2O_2 (Figure 2b). The former is the preferred approach when selecting a catalyst for the ORR due to its increased efficiency and the unstable nature of H_2O_2 compared to the stable nature of H_2O [23-28]. For the indirect pathway to occur, O_2 must first be adsorbed onto the surface of the catalyst. Thereafter, two H^+ additions take place, generating an adsorbed H_2O_2. The resultant H_2O_2 molecule can then go through more reduction to provide two H_2O molecules or dissociate, yielding a free H_2O_2. The direct reduction pathway is a dissociative adsorption process that starts with the O_2 adsorption on the surface of the catalyst. After the adsorption of O_2, the first e^- transfer occurs with the addition of H^+, producing *OOH. The second e^- transfer happens, this time in the form of another addition of hydrogen. The progress of the reaction depends on whether the second H^+ addition occurs at the O_2 molecule adsorbed to the EC or at the oxygen that was previously attached to the H^+. The EC will have two adsorbed OH groups in case the second H^+ addition occurs at the adsorbed oxygen to the EC (mechanism 1). At each *OH group, successive e^- transfers take place in the form of the addition of hydrogen, producing two molecules of H_2O molecules and O atoms that have been adsorbed desorb in case the second H^+ addition occurs at the oxygen that was previously attached to the first H^+ (mechanism 2). An H_2O water molecule desorbs after two subsequent transfers of e^- in the form of H^+ addition to the *O. According to Density functional theory (DFT) simulations of the two paths, the more plausible approach is mechanism 1, which is thermodynamically preferred [23-25].

Mechanism 1:

$O_2 \rightarrow {}^*O_2$
${}^*O_2 + e^- + H^+ \rightarrow {}^*OOH$
${}^*OOH + e^- + H^+ \rightarrow {}^*OH + {}^*OH$
$2{}^*OH + 2e^- + 2H^+ \rightarrow 2H_2O$
$O_2 \rightarrow {}^*O_2$

Mechanism 2:

${}^*O_2 + e^- + H^+ \rightarrow {}^*OOH$
${}^*OOH + e^- + H^+ \rightarrow {}^*O + H_2O$
${}^*O + e^- + H^+ \rightarrow {}^*OH$
${}^*OH + e^- + H^+ \rightarrow H_2O$

It is well understood that the ORR plays a vital role in numerous energy conversion devices. Still, the intrinsically sluggish nature, which is due to the huge overpotential of ORR, limits the overall efficiency of energy-conversion devices. Therefore, an appropriate EC is required that can effectively influence the kinetics of reactions and direct ORR towards more efficient pathways. As a result, developing novel cathode EC has recently become a key focus of industrial and academic research [23-29]. As per the DFT calculations, ORR catalytic performance correlates with the binding energy (BE) between oxygenated species and the surface of the catalyst. According to the Sabatier principle, a reactant should bind to a catalyst both firmly enough to cause a reaction to occur and weakly sufficient to cause the product to dissociate and prevent either the reactant or any intermediates from poisoning the catalyst surface. Therefore, in this instance, the oxygen intermediate should be bound to the EC firmly enough for the reaction to occur on the interface but not too strong that oxygenated species contaminate the EC. It is possible to develop "volcano plots" like the one shown in Figure 3, correlating catalytic activity to the BE of oxygen species using experimental data and numerical simulations. To determine the ideal BE for ORR ECs, these plots can then be analyzed. Pt ECs hold the ideal position among the ECs considered in Figure 3; however, it is not at the volcano's highest position [23, 28]. As a result, constructing an ORR ECs still has space for development. One way to alter the adsorption energies of oxygen intermediates is to modify the catalyst's conductive band or d-band center. The adsorption energies have been demonstrated to correspond with the d-band, which represents the electronic properties of the solid. It has been found that the d-band center correlates with adsorption energies and serves as a descriptor of the solid's electronic characteristics [23, 28]. It has been observed that in bimetallic ECs, tuning the morphology and predominant facets can change the d-band center's position in relation to bulk metals.

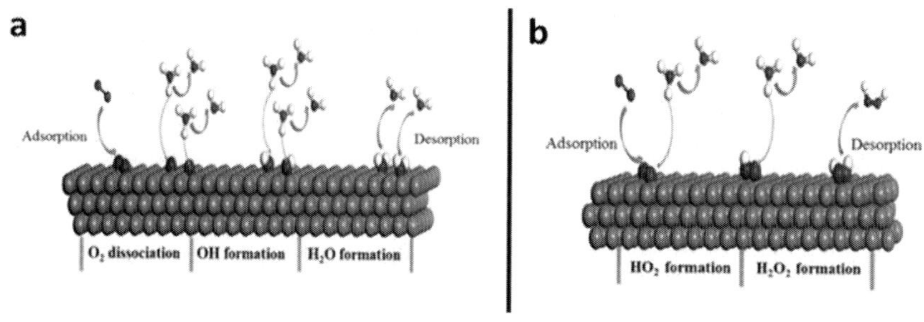

Figure 2. Proposed mechanism schematics of (a) full reduction and (b) partial reduction of oxygen, Adapted with permission from Reference [24], Copyright (2018), Multidisciplinary Digital Publishing Institute.

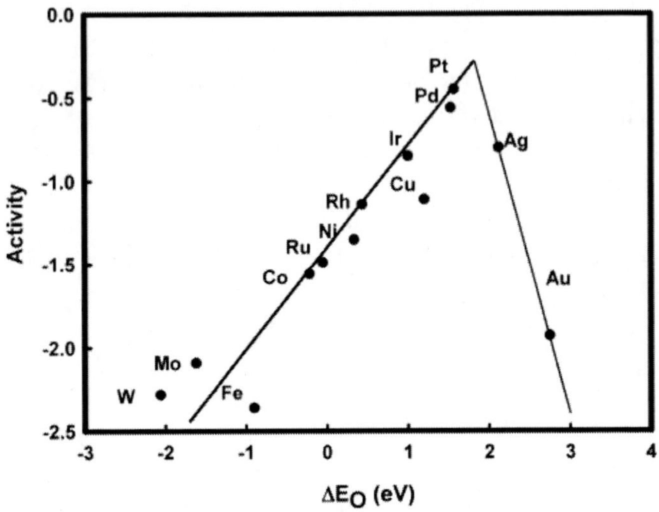

Figure 3. Trends in oxygen reduction activity (defined in the text) plotted as a function of the oxygen binding energy. Adapted with permission from Reference [28], Copyright (2004), American Chemical Society.

2.1. Pt-Based Bimetallic ECs

Pt stands out as the ideal catalyst among present pure metals because it possesses all the necessary properties for an effective ORR. The slow ORR kinetics, however, requires a more significant loading of the Pt EC. Since Pt is a noble metal in and of itself, the limited supply of Pt, CO poisoning, durability concerns, and its high price emerge as further obstacles. Overall, the limitations mentioned above prevent the widespread practical use of FCs, even though alkaline FCs using Pt as an ORR EC have been successfully constructed for the Gemini and Apollo spacecraft flown by NASA in the 1960s [26-27]. Most current ORR works are focused on ways to lower costs by reducing cathode Pt loadings or replacing Pt with other materials while maintaining high performance. The development of inexpensive, highly effective ORR ECs has recently focused on alloying Pt with other non-Pt metals to reduce the Pt content. In this respect, Pt-based bimetallic alloys (nanoparticles) NPs with cost-effective transition metals (M) have been suggested to replace conventional catalytic systems for various applications. Incorporating another metal reduces Pt loading while maintaining high activity and mass activity (MA). Another reason to develop Pt-based bimetallic alloy ECs is their better catalytic performance than pure Pt ECs. Incorporating other metals in the Pt ECs tunes its surface electronic structure and reactivity through ligand, strain, and ensemble effects. The structure of Pt-based bimetallic alloy ECs significantly impacts their electrocatalytic activity [27, 29]. They offer enormous opportunities to enhance the catalytic activity of bimetallic nanocatalysts by altering their morphology and hybrid nanoscale properties. It has been observed that Pt-alloy ECs that bind to oxygen intermediates (such as *OH, where * designates an adsorption site on the catalyst's surface) are relatively weaker than pure Pt for the ORR and thus enhance the kinetics. Hao and coworkers synthesized PtCo NPs anchored on 3D-hierarchical porous graphene by introducing oxygen functional groups (PtCo/3D rHPGO) by acid treatment method (Figure 4a). To compare the electrochemical activity of as-synthesized NPs, they have also synthesized PtCo supported on multiwalled carbon nanotubes (MWCNT) and 3D-hierarchical porous graphene (3D HPG) labeled as Pt-Co/MWCNT and Pt-Co/3D HPG NPs. A comparison was made between the Pt-Co/3D rHPGO EC's initial activity and durability and that of the standard Pt/C. The CV and LSV plots of Pt-Co/3D rHPGO maintain a similar performance to the initial after 20000 ADT cycles (Figure 4b and c).

However, the Pt/C-JM catalyst's activity drastically decreased, with a 65.4% decrease in ECSA (Figure 4d) and a 52.7% decrease in MA (Figure 4e). As a result, Pt-Co/3D rHPGO outperformed Pt/C in all aspects, including initial activities and durability. This was owing to the presence of 3D rHPGO, which provides spatial protection to the Pt-Co NPs and the presence of oxygen atoms with an anchoring effect [30]. Kim et al. designed atomically ordered PtCu supported on carbon (O-PtCuNF/C). The intermetallic PtCu L11 composition and the benefits of nanoframes are combined in the O-PtCuNF/C to produce higher ORR performance compared to disordered alloy PtCu nanoframes (D-PtCuNF/C) and conventional Pt/C ECs. The O-PtCuNF/C also displayed the most excellent ORR MA among the PtCu-based catalysts. Additionally, compared to D-PtCuNF/C and Pt/C, the O-PtCuNF/C demonstrated better ORR durability and significantly less etching of constituent atoms, showcasing the chemical stability of intermetallic structure [31]. Li and coworkers used a one-step modified glycol reduction method to load PtFe alloy NPs onto porous carbon nanofiber (PCNF) while maintaining the pH of the solution. The PtFe alloy NPs with a confined particle size distribution can be distributed evenly on the surface of PCNF. The catalyst $Pt_{4.8}Fe$/PCNF with PCNF as carbon matrix demonstrates improved ORR activity and has the earliest onset potential (E_{onset}) than commercial Pt/C.

Additionally, accelerated durability testing (ADT) performed after 2000 cycles reveals comparable durability. The electronic properties of the Fe metal in the PtFe alloy may account for the remarkable performance of $Pt_{4.8}Fe$/PCNF EC. PCNF is also expected to boost performance thanks to its rough surface and porous structure [32].

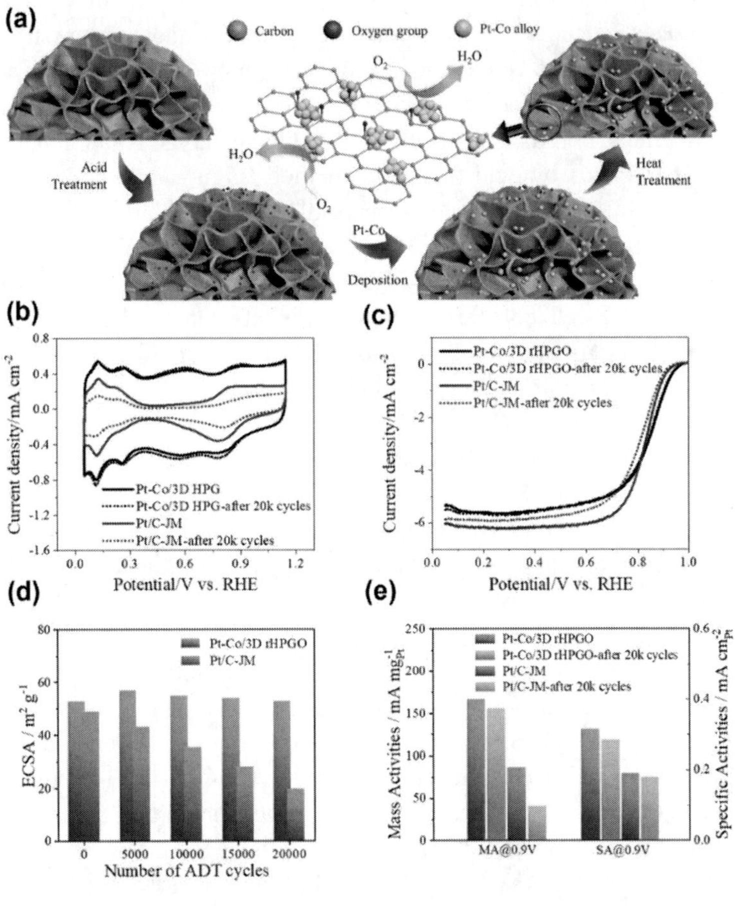

Figure 4. (a) Illustration of the synthesis of the Pt-Co/3D rHPGO catalyst. Comparison of Pt-Co/3D rHPGO and Pt/C-JM during ADT: (b) CV curves, (c) LSV curves, (d) normalized ECSA, and (e) MA and SA at 0.9 V. Adapted with permission from Reference [30], Copyright (2021), American Chemical Society.

In recent years, morphology-controlled Pt-based catalysts have drastically enhanced the activity of Pt nanocrystals (NCs) and thus provide new insight into the development of ECs for ORR. A recent study by Zhu and co-workers demonstrated that concave octahedron Pt–Ni/C catalyst exhibited excellent ORR activity and enhanced durability than conventional Pt/C. The MA and specific activity (SA) of concave octahedron Pt–Ni/C ECs reached 1.018 Amg^{-1} and 1.63 $mAcm^{-2}$, which were around 13.8 and 12.5 times higher than the commercial Pt/C. The enhanced performance of these ECs indicates that their concave structure improves active sites and enhances the utilization of Pt [33]. However, Guo et al. proposed that the PtSn nanowire (NW) is more active than Pt NWs and PtSn NPs based on their experimental and theoretical calculations. The improved performance can be ascribed to the effects of Sn and its oxides on the electronic structures and structure that decrease the oxophilic nature and the d-band center of the Pt surface and thus increase the stability during the durability measurements [34]. Lai et al. reported the synthesis of surfactant free PtAg fractal dendritic NCs with a large surface area which were synthesized via a facile one-pot reaction strategy. The as-prepared NCs displayed higher electrochemical catalytic activities and structural stabilities than standard Pt/C towards ORR in an acidic medium because of their unique structural feature and clean surface. The proposed method may be advantageous when developing highly active and durable Pt-M ECs for electrocatalytic applications [35]. These studies showed that ORR activity is strongly dependent on morphology engineering.

2.2. Pd-Based Bimetallic ECs

Pd-based ECs have received a lot of interest owing to their analogous physical structures, including the fcc crystal structure and a similar atomic radius, comparable performance, and lower price than Pt. Extensive research has shown that pure Pd and Pd-based ECs have higher catalytic activity and stability than Pt-based ones. Furthermore, it was demonstrated that Pd-based ECs show significantly greater methanol tolerance and resistance to CO compared to Pt-based electrocatalysts [29]. Therefore, it is highly desired and necessary to create high-efficiency and cost-effective Pd-based ECs, for significantly improving ORR performance. While ORR ECs are projected to have the qualities of low cost, high activity, and high stability synchronously, they still do not meet all the requirements for implementing FCs. To achieve this, several techniques have been developed for further enhancing the ORR performance of Pd-based ECs with fine control over the shape, composition, crystal plane, and surface strain while observably reducing the Pd usage and modifying the electronic structure [29, 36-41].

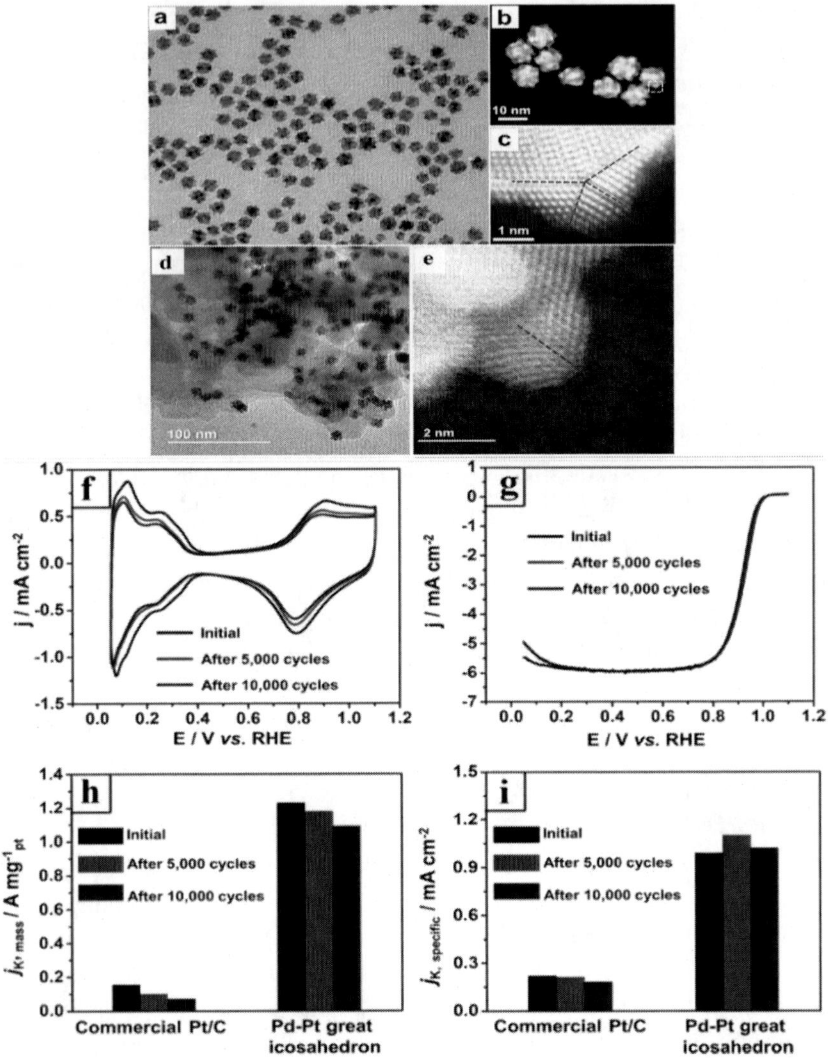

Figure 5. (a) TEM and (b) HAADF-STEM images of the as-obtained Pd−Pt great icosahedra. (c) Atomic-resolution STEM image taken from the corner region of a great icosahedron, as marked by the box in panel (b) The twin boundaries are labeled with red dashed lines (d) TEM and (e) HAADF-STEM images of the carbon-supported Pd−Pt great icosahedra after 10000 cycles of ADT. Benchmarking of the Pd−Pt great icosahedra against the commercial Pt/C catalyst. (f) CV and (g) oxygen reduction polarization curves recorded from the Pd−Pt great icosahedra before and after 5000 and 10000 cycles of ADT. (h) Mass and (i) specific activities of the catalysts at 0.9 V vs RHE before and after ADT. Adapted with permission from Reference [41], Copyright (2021), American Chemical Society.

In recent years, numerous Pd-based alloy NPs with different compositions have been explored. As a result, increased ORR performances of these alloy systems have been observed. We synthesized PdNi alloy system and studied its ORR performance. As-prepared Pd_3Ni/C NPs are superior to the commercial Pt/C and Pd/C ECs. The improved performance of Pd_3Ni/C is primarily due to the downshift of Pd d-band center, larger ECSA, and uniform distribution of the PdNi alloy NPs on the carbon matrix. The computational analyses also demonstrated the downshift of the Pd d-band center for Pd_3Ni/C, which makes it superior to all other ECs [36]. Jaramillo and co-workers reported the electron-beam PVD synthesis method to synthesize $Ag_{1-x}Pd_x$ thin films with $0 \leq x \leq 1$. The designed nanomaterials and nanofilms remained approximately constant after electrochemical measurements, inferred from XRD and XPS analysis. Based on experimental findings and DFT simulations, the observed ORR activity improvements are mainly ascribed to electronically increased Pd active sites by introducing Ag atoms within the lattice via ligand or strain factors that result in enhanced oxygen BE [37]. In another work, Zhang et al. reported the PdCo encapsulated in N-doped porous carbon (PdCo@NPNC) synthesis and studied their catalytic behavior toward ORR. The carbon shell prevents the NPs from coagulation, detachment and dissolution while the FC operates. The optimized PdCo@NPNCs outperforms the state-of-the-art Pt/C and Pd/C catalysts in electrocatalytic performance and enhanced durability for ORR, owing to their unique structures and chemical compositions [38].

Significant advancements have been made in recent years in preparation techniques for various morphology-controlled NPs. These size and shape-controlled NPs replicate the catalytic activity of the bulk metal, while they exhibit more catalytic activity because there are more active sites. To this end, Wang et al. Pd nanobelts (NB) catalysts with expanded lattice were designed by a simple solvothermal method. It was found that as-synthesized Pd-nanobelts possess extended lattices that shift the d-band center positively. The interaction of oxygen 2p states with the metal d states may be enhanced by an upshifting in the d-band center. The additional adsorbate onto the Pd-surface is also altered by a change in the d-center, which improves the catalytic activity and durability [39]. Similarly, Li and coworkers reported the synthesis of Pd with highly wrinkled nanosheets (HW-PdNs). Compared to standard Pd nanosheets (PdNs), commercial Pd/C, and Pt/C, the HW-PdNs showed significantly increased catalytic performance. The high catalytic performance of the ECs could be mainly ascribed to the compressive strain created from the wrinkle nanostructure, which reduces the binding strength of adsorbed oxygen-intermediate and results in fast kinetics toward ORR. Additionally, the increased stability of HW PdNs is attributable to the altered electronic structure of Pd resulting from the formation of wrinkles [40]. It was reported that Pt nucleation and growth were confined to the vertex of Pd icosahedral seeds, resulting in Pd–Pt icosahedra with multiple twinned Pt dots. Twin defects can also boost the ORR performance of Pt NCs by utilizing the surface strains generated by the defects. Figure 5 (a-c) represents the TEM and HAADF-STEM images of the Pd−Pt icosahedra. The ORR activity and durability of the Pt dots are much higher than the state-of-the-art Pt/C because of their highly twinned shape and robust anchoring to the surface of Pd. An ADT was also used to investigate and compare the stability of the Pd–Pt icosahedra to that of Pt/C. CV and LSV plots were plotted after 5000 and 10000 ADT cycles and a minimal change in peak potential and J_k was observed, as presented in Figure 5f and g, respectively. After 10,000 redox cycles, a rise in the MA (Figure 5h) and the SA (Figure 5i) of the Pd–Pt icosahedra were observed compared to the pure Pt/C. TEM and HAADF-STEM characterization of the Pd−Pt icosahedra was done after 10000 ADT cycles to understand the durability improvement (Figures 5d and e). It is observed that on the carbon support, the catalytic NPs were still evenly distributed and had barely any agglomeration or detachment. The improved stability of as-synthesized catalysts was facilitated by their high specific surface area and concave structure, which allowed for additional anchoring sites on the carbon support [41].

2.3. Transition Metal-Based Bimetallic ECs

The lack of availability in nature, high price, poor stability, and occasionally electrode poisoning remain problems for platinum group metal catalysts (PGM-Catalysts). To address these problems, non-Pt group metals (non-PGM) ECs are now being used as alternatives in developing more effective and long-lasting ORR ECs. Different research groups have studied non-PGM (Co, Fe, Mn, Mo, etc.) based ECs. Despite frequently having problems with low conductivity and operating-condition stability, ORR ECs made from non-PGMs have usually performed exceptionally well in the half-cell reaction. Conductivity has been significantly increased by supporting non-PGMs across the top of the carbon matrix with adequate anchoring sites [42, 43]. In Kurungot and workers' report, a facile and efficient microwave irradiation process was used to deposit CoMn alloy oxide NPs over porous graphene. The

Co:Mn ratio was varied to develop a variety of catalysts, but the catalyst with a 2:1 Co:Mn ratio designated CoMn/pNGr (2:1) exhibited noticeably better activity toward ORR in 0.1 M KOH electrolyte. It displayed a potential shift of around 60 mV with a smaller Tafel slope of 74 $mVdec^{-1}$, similar to that obtained from the conventional Pt/C catalyst. This remarkable ORR performance of CoMn/pNGr (2:1) has been attributed to a synergistic interaction between the metal particles and the defects in the porous N-doped graphene [44]. In another work, an ultrasonic-assisted low-temperature chemical reduction method was adopted to develop a novel Ag_4Sn/C intermetallic catalyst. The Ag_4Sn intermetallic catalyst displays improved electrocatalytic activity and follow $4e^-$ pathway towards ORR. Ag_4Sn/C also showed great methanol tolerance and outstanding long-term stability in an alkaline electrolyte [45]. Nguyen et al. synthesized binary CuFe alloy NPs firmly enclosed within nitrogen-doped graphene nanosheets (CuFe NPs/NrGO). The hybrid NPs showed well ORR performance in 0.1M KOH solution. Additionally, the catalyst outperformed commercial Pt/C products regarding stability and methanol tolerance. The improved ORR performance of the CuFe NPs/N-rGO results from an effective synergistic interaction between CuFe NPs and N-rGO nanosheets. Remarkably, the development of an encapsulation structure, in which many layers of N-rGO enclose CuFe NPs, not only enhances e^- transport but also successfully prevents the dissolution and agglomeration of the active sites of metal NPs, enhancing the catalytic performance, reaction kinetics, and stability [46].

In the previous section, we discussed the advantages of nanostructured materials and their role in enhancing catalytic activity in energy storage devices. Nanosponges have a peculiar 3D architecture that offers a significant surface defect and excellent structural stability, resulting in increased catalytic activity. Moreover, nanosponges have significantly enhanced electrode efficiency due to their vast conducting surface, electronic properties, and strong crystalline phase stability. For instance, Wang et al. reported a typical kinetically controlled solution strategy to synthesize Au−Ag nanoporous sponges (NSs) comprised of 3D networks without using any surfactants. The distinct AuAg NSs successfully incorporate the strong electronic effect, interconnected porous architecture, and self-supporting structure. These essential factors allow them to display increased ORR catalysis and exceptional stability in alkaline electrolyte. The obtained $AuAg_{3.2}$ NSs exhibit remarkable stability with minimal activity decay after 10,000 redox cycles and offer a MA of 1.26 A mg_{Au}^{-1}, which is about 8.2 times higher than the commercial Pt/C [47]. To improve ORR performance, materials scientists are working to identify several types of catalyst carriers. Researchers started their exploration with functionalized and heteroatom-doped carbon support in this case. In a recent study, Li and co-workers constructed Fe-Co dual sites incorporated on N-doped porous carbon and evaluated its ORR performance, as presented in Figure 6a. TEM (HRTEM) and HAADF-STEM revealed the uniform Fe and Co atoms distribution over a porous carbon matrix (Figure 6b-d). The comparison of all the EC in terms of $E_{1/2}$, E_{onset} and Tafel slope are presented in Figure 6e-g. The catalyst displayed a superior ORR performance, with the earliest E_{onset} of 1.06 V (vs RHE) and $E_{1/2}$ of 0.863 V (vs RHE) and a lower Tafel slope than benchmark Pt/C. After 50,000 electrode measurement cycles and 100 h of operation in an H_2/air single cell, the catalyst with dual metal sites is observed to be remarkably stable (Figure 6i). It displayed a low yield of H_2O_2 production and lower tolerance to CO and CH_3OH (Figure 6h and j). According to DFT calculations, the activation of O-O, which is essential for the $4e^-$ ORR process, is more likely to occur at dual sites [48].

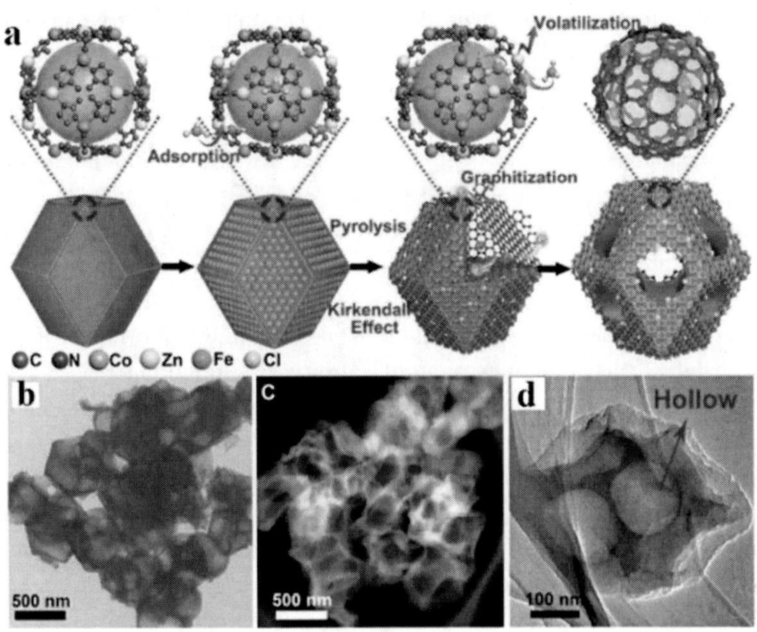

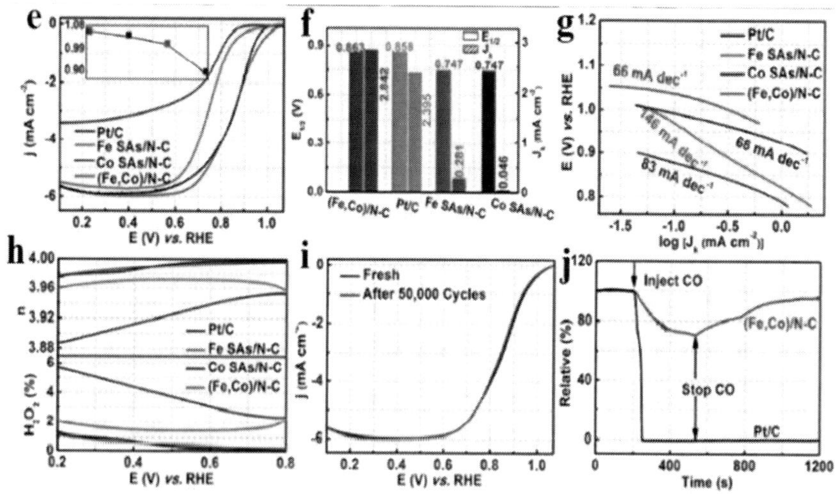

Figure 6. (a) Preparation of (Fe,Co)/N-C. (b) TEM, (c) HAADF-STEM, (d) HRTEM of (Fe,Co)/N-C. (e) RDE polarization curves of Pt/C, Co SAs/N-C, Fe SAs/NC, and (Fe,Co)/N-C in O_2-saturated 0.1 M $HClO_4$ with sweep rate 10 mV s^{-1} and rotation rate 1600 rpm. Inset: E_{onset} of different catalysts. (f) Comparison of $E_{1/2}$ and J_K of different catalysts. (g) Corresponding Tafel plots from RDE polarization curves. (h) Electron transfer number (n) (top) and H_2O_2 yield (bottom) vs potential. (i) RDE polarization curves of (Fe,Co)/N-C before and after 50,000 potential cycles ranging from 0.6 to 1.0 V (vs RHE) in O_2-saturated 0.1M $HClO_4$ (j) Tolerance to carbon monoxide of (Fe,Co)/N-C compared with Pt/C (20%) electrocatalyst at 0.5 V (vs RHE). CO (flow rate: 100 mL s^{-1}) is injected into electrolyte (100 mL) at 200 s and stopped at 500 s. Adapted with permission from Reference [48], Copyright (2021), American Chemical Society.

In another work, Ni/Cu incorporated N-doped porous carbon materials ($Ni_{1-x}Cu_x$/N C) with varied Ni/Cu ratios were reported. It was observed that the Ni/Cu ratio was found to have a substantial impact on the $Ni_{1-x}Cu_x$ particle sizes, specific surface area, morphology, and the porous nature of the ECs, ultimately altering the ORR performance. Among the investigated $Ni_{1-x}Cu_x$/N C catalysts, $Ni_{0.7}Cu_{0.3}$/N-C showed the best ORR activity with an $E_{1/2}$ of 0.783 V (vs RHE) and a j_m of 4.85 mAcm^{-2} in alkaline electrolyte. DFT simulations show that the synergistic factor of bimetallic Ni/Cu can promote the adsorption/desorption properties of the intermediate and accelerate the ORR performance [49].

3. Urea Oxidation Reaction

In the previous section, we have already come across the fact about how non-conventional energy conversion systems have become the need of the hour. Also, H_2 fuel is considered the most promising candidate to replace conventional fuel resources due to its high energy density and environmentally safe properties. However, the promotion of H_2 FC is still facing various issues due to some critical problems related to expenses and efficiency in the present stage of research in order to make it commercially available. Here, the inclusion of urea as a 'clean' and 'advanced' fuel option may not align with the idea of a "clean energy source," primarily because the main source of urea is human and animal urine. However, in reality, it serves the concept of a sustainable energy source quite effectively. A substantial amount of untreated wastewater from industries, human as well as animal urine are released into the environment without being treated which causes both water and soil contamination which further leads to health and environmental hazards [50]. With the advancement of technology and hardwork of researchers, it has been made possible to replace the previously used traditional cost ineffective and unsuitable methods for urea treatment with urea based energy conversion techniques such as photoelectrochemical urea splitting, urea electrolysis, DUFC, etc. These methods are found to be more advantageous over H_2 based energy conversion technologies:

1. Urea being naturally abundant, is more safer since it is non-flammable apart from being easy for transportation and also provides with theoretical open circuit voltage of 1.15V in DUFCs which is similar to that of H_2 FC (1.27 V vs. RHE) [51,52].
2. Usage of urea as a fuel serves the purpose of both wastewater treatment as well as clean energy resource [53].
3. Theoretical cell voltage required for urea splitting is 0.37 V vs. RHE which is far below than water splitting i.e 1.23 V vs. RHE and since urea consists of 6.67 wt% of hydrogen, it is also considered to be potential hydrogen storage chemical [54].

The underlying reaction in all the methods discussed above involves the UOR which is based on a 6 e^- transfer process. Inspite of so many merits of urea based technologies, the UOR needs to be optimised in presence of a cost effective EC which works as a means to lower the potential barriers and improve the reaction kinetics in order to be able for commercialization [55].

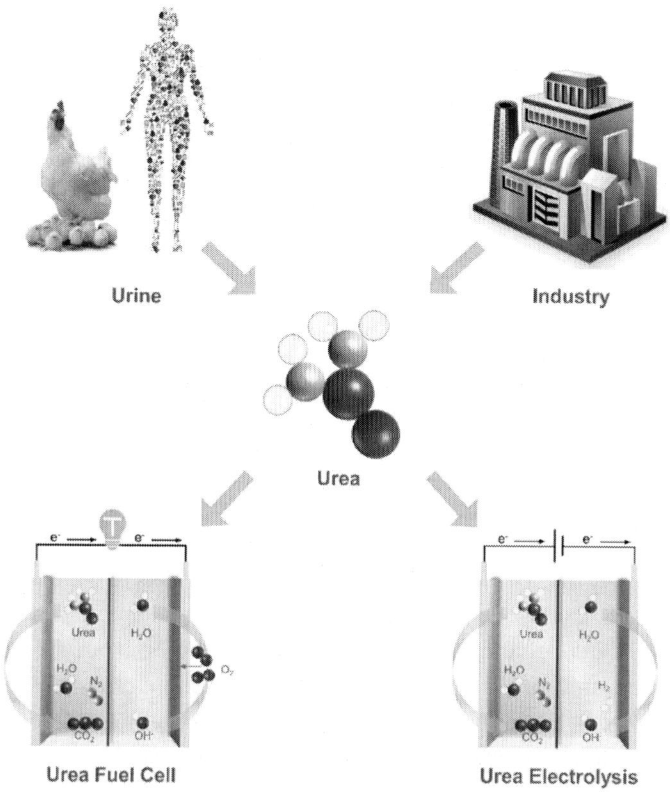

Scheme 1. An illustration of urea-based energy conversion systems, including urea electrolysis and urea FC. Adapted with permission from Reference [21], Copyright (2022), Multidisciplinary Digital Publishing Institute.

The UOR is most favourably carried out in alkaline medium in presence of the electrocatalyst and the reactions involved in urea electrolysis are given as below [50]:

Anodic reaction: $CO(NH_2)_2(aq) + 6OH^- \rightarrow N_2(g) + 5H_2O(l) + CO_2(g) + 6e^-$

Cathodic reaction: $6H_2O(l) + 6e^- \rightarrow 3H_2(g) + 6OH^-(aq)$

Overall reaction: $CO(NH_2)_2(aq) + H_2O(l) \rightarrow 3H_2(g) + N_2(g) + CO_2(g)$

The best part about UOR unlike ORR is that it can be achieved even without the inclusion of noble metals under alkaline medium. However, Pt was the first EC used for urea oxidation for medical treatments of artificial kidneys, and in biochemical sensors during 1970's [56, 57]. The transition metal so far which has been mostly used for its catalytic property towards electrooxidation of urea is Ni, although nowadays transition metal such as Co, Fe, Cu and Mn are also being studied for UOR. Thus, in this chapter the UOR ECs have been broadly classified as Ni-based and Ni-free catalysts.

3.1. Ni-Based Bimetallic ECs

The idea of Ni being an active metal towards urea oxidation came from urease, a cytosolic enzyme which catalyses the hydrolysis of urea to ammonia and CO_2. Today, Ni- based catalysts have been luckily found to show promising catalytic activity towards UOR which surpasses the problem of its slow reaction kinetics by assisting the reaction without any cost and reserve restriction unlike nobel metals such as Pt, Pd [58]. Further, due to the high stability in alkaline media and easy formation of its active phase towards UOR from the multiple oxidation states, Ni-based ECs have gained increasing attention. In 2009, Boggs et al. developed a technology where they accomplished collection of H_2 from the cathode and N_2 from the anode using a Hoffmann apparatus by direct conversion of urine and urea by applying a cell potential of 1.4 V vs. RHE with the assist of a non-precious Ni catalyst in alkaline medium. They concluded that human urine with a minimum concentration of 0.33 M can be converted to N_2, H_2 and CO_2 via plausible mechanism [57]:

Anodic reaction: $CO(NH_2)_2(aq) + 6OH^- \rightarrow N_2(g) + 5H_2O(l) + CO_2(g) + 6e^-$

$Ni(OH)_2(s) + OH^- \rightarrow NiOOH(s) + H_2O(l) + e^-$

Cathodic reaction: $6H_2O(l) + 6e^- \rightarrow 3H_2(g) + 6OH^-$

Overall reaction: $CO(NH_2)_2(aq) + H_2O(l) \rightarrow N_2(g) + 3H_2(g) + CO_2(g)$

where 2nd equation is considered to be the attributing factor towards the current.

Further, Vedharathinam and Botte gave significant contribution towards the understanding of mechanism of electrooxidation of urea on a Ni-catalyst proposed by the previous authors through the study of electrochemical and in situ techniques where they confirmed that Ni^{3+} was the active specimen for the UOR, and suggested a mixed reaction mechanism controlled by both electrode kinetics and diffusion. With the help of in situ surface-enhanced Raman spectroscopy they proved that urea oxidation on NiOOH follows a catalyst regeneration mechanism [58, 59]. Any Ni-based EC is known to be pre-catalyst, as in order to become UOR active, they are converted into $Ni(OH)_2$ in strong alkaline medium leading to the formation of Ni^{3+} [60]. From DFT analysis Daramola et al. found that the rate determining step was the step in which CO_2 gets desorbed from the surface of the catalyst [61]. The whole process of catalytic reaction occurs through the molecules getting adsorbed, then the chemical reaction occurs and lastly desorption occurs from the surface of the catalyst. Hence, large availability of surface area of the catalyst for the molecules and its good stability is a must. However, due to intermediate poisoning, the availability of catalyst's surface area decreases, thus resulting in reduction of efficiency. Such drawbacks of various monometallic ECs such as intermediate poisoning, low stability, high over potential, etc. are lessened or diminished by inclusion of another metal which complements the former complex via synergistic effect.

Munde et al. synthesised Ni-Bi NPs where he concluded that as synthesised NPs showed 8.3 times more enhancement factor when compared to Ni particles and explained the reason for such improved character due to the presence of Bi, which pulls the electron density of Ni and makes the surface electron deficient, this results in faster adsorption of OH^-, further leading to faster electrokinetics towards Ni-OOH formation. Munde also pointed out that Bi NPs were also involved in the molecular oxygen stabilization step, and all of these factors resulted in increased charge transfer and electrochemical activity at a very low overpotential towards UOR [62].

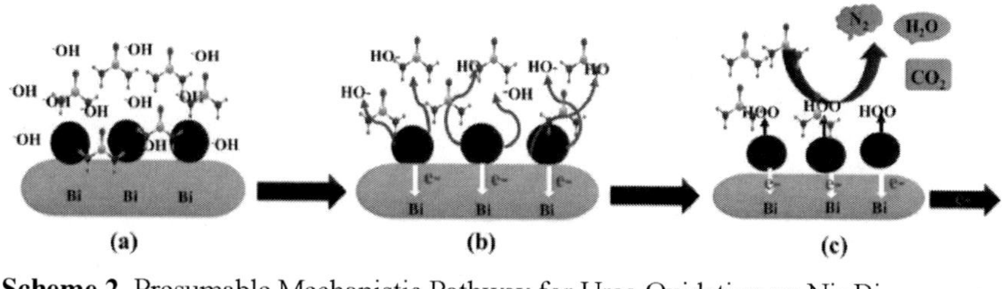

Scheme 2. Presumable Mechanistic Pathway for Urea Oxidation on Ni–Bi Bimetallic NPs Corresponding to (a) Ni–Bi–OH and Ni–Bi–O–, (b) Bi–Ni–O–OH, and (c) Conversion of Urea to N_2, H_2, and CO_2. Adapted with permission from Reference [62], Copyright (2021), American Chemical Society.

Yang et al. explored the doping of $Ni(OH)_2$ with Mn with varying composition ratio and found that Mn doping facilitates the conversion of Ni^{2+} to Ni^{3+} assisting the reversibility of $Ni(OH)_2$/NiOOH which results in lowered overpotential, which means better UOR performance. Through Raman spectroscopy it was concluded that Mn is involved in increasing the bond length of Ni-O which leads to the phase structure distortion resulting in enhanced catalytic activity of Ni_{1-x} Mn_x layered double hydroxide (LDH) when compared to $Ni(OH)_2$[63]. Another problem faced by the ECs during catalysis of UOR mentioned earlier is its intermediate poisoning. As reported by Daramola et al. CO groups limit the catalytic activity by strongly adsorbing on the Ni active sites, resulting in lesser available active sites for UOR. To address this drawback, King et al. did an investigation based on multi-metal catalysts for stable UOR where they found that the combination of Ni and Rh gives the best synergistic effect (among Ni-Pt, Ni-Ir, Ni-Ru and Ni-Rh) where Rh reduces the surface poisoning and increases the catalytic stability [64].

In an overall water splitting reaction, the high theoretical thermodynamic voltage is mostly caused due to the more sluggish anodic oxygen evolution reaction (OER) compared to the cathodic one. Hence, one of the ways adapted to increase the energy conversion efficiency is by replacing OER by UOR. In the past few years, Ni-based spinels with formula AB_2X_4 (where A and B are tetrahedrally and octahedrally coordinated cations, and X is an anion) have gained attention towards electrocatalytically aiding OER. Since replacing OER by UOR is already a strategy for energy conversion systems, Ni- based spinels can also be used for UOR where Ni can provide the active sites. Liang and co-workers proposed Ni present in urchin-like $Ni_xCo_{3-x}O_4$ (x=0, 0.75, 0.86,1 and 1.20), served as an active phase toward UOR catalysis. They found that by increasing the number of Ni ions into the octahedral sites, an increase in the number of Ni redox equivalents results; which promote the performance of UOR [65].

Figure 7a depicts how in the assynthesised inverse spinel, Ni ions are only occupying the octahedral sites and some Co^{3+} also occupy the octahedral positions, and the remaining Co_{1-x} $^{2+}$ and Co_x $^{3+}$ take up the tetrahedral positions. And the study of morphology was done by SEM (Figure 7b) where the catalyst displayed 3D urchin-like structure. EDS elemental mapping confirmed uniform distribution of Ni, Co and O throughout the catalyst in Figure 7c. TEM images of the ECs in Figure 7d confirmed the formation of nanowires made up of NPs. HRTEM image given in Figure 7e displays clear and ordered lattice formation of the complex which suggests good crystallinity with lattice spacings of 0.203, 0.245, and 0.287 nm, ascribed to the (400), (220), and (311) planes of $NiCo_2O_4$, respectively. Also the polycrystallinity of $NiCo_2O_4$ spinel is confirmed by the diffraction rings in the SAED pattern shown in Figure 7f. The electrochemical measurements of OER and UOR were carried out by performing CV in a scan rate of 100 m Vs^{-1} in 1M KOH with and without 0.33 M urea. Redox peaks for Co_3O_4 catalyst are around 1.45

V vs. RHE (corresponding to Co^{3+}/Co^{4+} redox transition), whereas for Ni containing catalysts there is a shift of potential towards 1.29 V vs. RHE in absence of urea. The inset graph in Figure 7g suggest how incorporation of Ni increases the OER activity when compared to Co_3O_4.

However, there is a decrease in OER performance with increase in x in $Ni_xCo_{3-x}O_4$, which suggests us that Co is more active towards OER, thus decrease in its amount in the catalyst reduces its activity towards OER. However due to synergistic effect $Ni_{0.75}Co_{2.25}O_4$ shows the best performance out of the five. Figure 7h shows the urea oxidation polarization curves where at 1.3 V vs. RHE corresponds to the formation of NiOOH and at a potential of 1.45 vs. RHE there is a sharp increase in anodic current density attributing to the UOR due to the interaction of urea molecules at the Ni^{3+} activated sites. Here, the maximum and minimum current density corresponds to the $Ni_{1.20}Co_{1.80}O_4$ and Co_3O_4 and also the onset potential is negatively shifted with increase in Ni content, suggesting Ni to be responsible for enhanced UOR performance. Figure 7i-j corresponds to the turnover frequency (TOF) of only the Ni containing catalysts in KOH in absence (Figure 7i) and presence (Figure 7j) of urea, repectively. TOF value of $Ni_{1.20}Co_{1.80}O_4$ is found to be highest in presence of urea, whereas TOF value for $Ni_{0.75}Co_{2.25}O_4$ is found to be highest in absence of urea (i.e for OER). The kinetics of UOR was determined by the Tafel slope distribution as shown in figure 7k, where $Ni_{1.20}Co_{1.80}O_4$ showed the fastest reaction rate due to the octahedral sites being mostly covered by Ni ions. Figure 7l displays the potential of the ECs in 1M KOH to reach a value of $j = 10$ mA cm^{-2} towards OER (absence of urea) and UOR (presence of urea), where it has been concluded that in $Ni_xCo_{3-x}O_4$ with increase in value of x, the potential at a j of 10 mA cm^{-2} for the OER increases and the trend goes opposite for UOR [65].

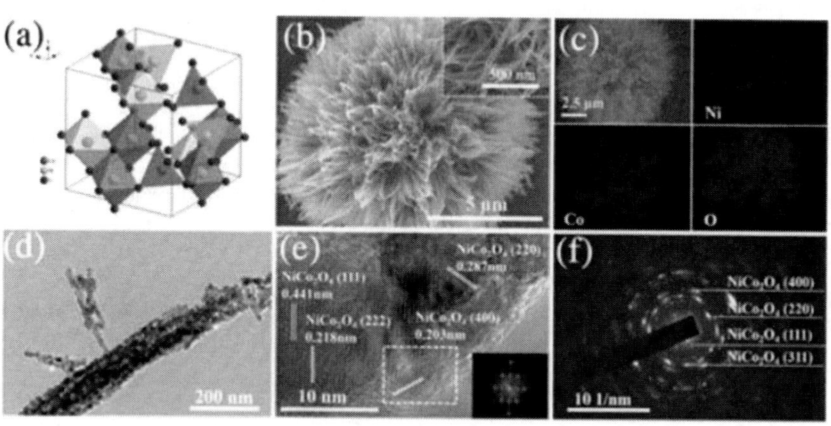

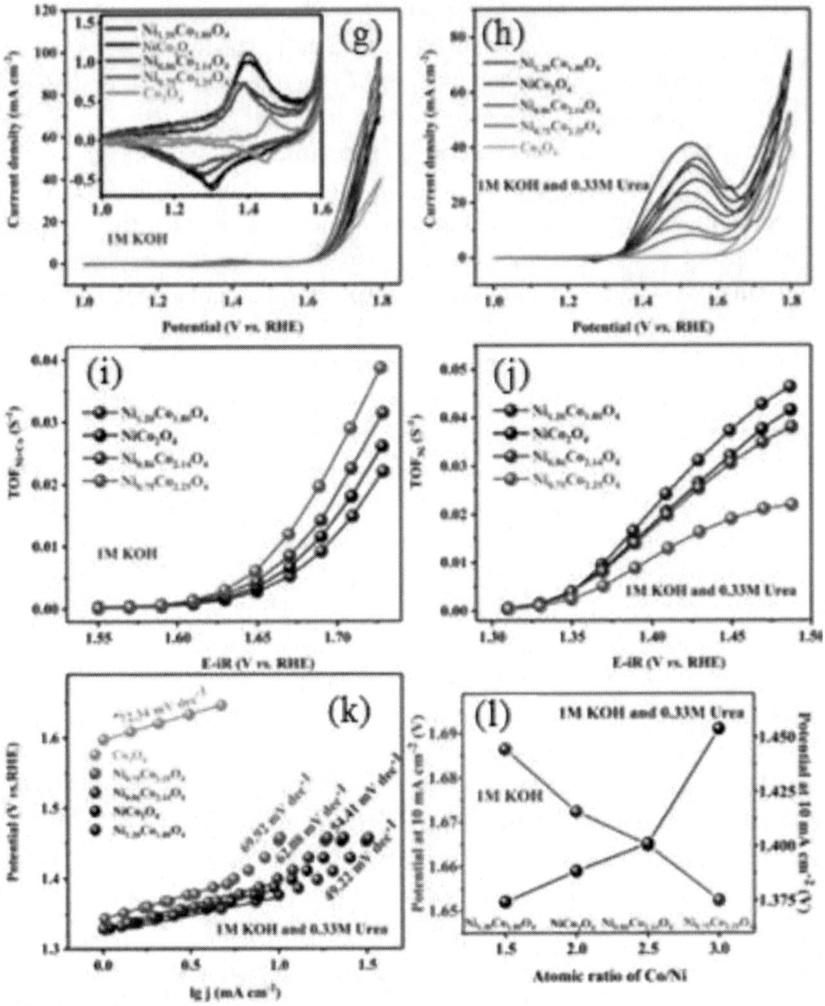

Figure 7. (a) Crystal structure of $Ni_xCo_{3-x}O_4$ (x = 0, 0.75, 0.86, 1.00, and 1.20). (b) 3D urchin-like structure, (c) EDS element mapping, (d) TEM image, (e) High-resolution TEM image and (f) SAED pattern of $NiCo_2O_4$ spinel. Cyclic voltammograms (CVs.) were recorded after CV activation for 100 cycles at a scan rate of 100 mV s^{-1} in 1 M KOH in the (g) absence and (h) presence of 0.33 M urea in 1 M KOH at a rate of 10 mV s^{-1} . Turnover frequency (TOF) of the catalysts in the (i) absence and (j) presence of 0.33 M urea in 1 M KOH. (k) Tafel slope after IR compensation at a rate of 2 mV s^{-1}. (l) Potentials of the OER and UOR at different metal atomic ratios. Adapted with permission from Reference [65], Copyright (2021), American Chemical Society.

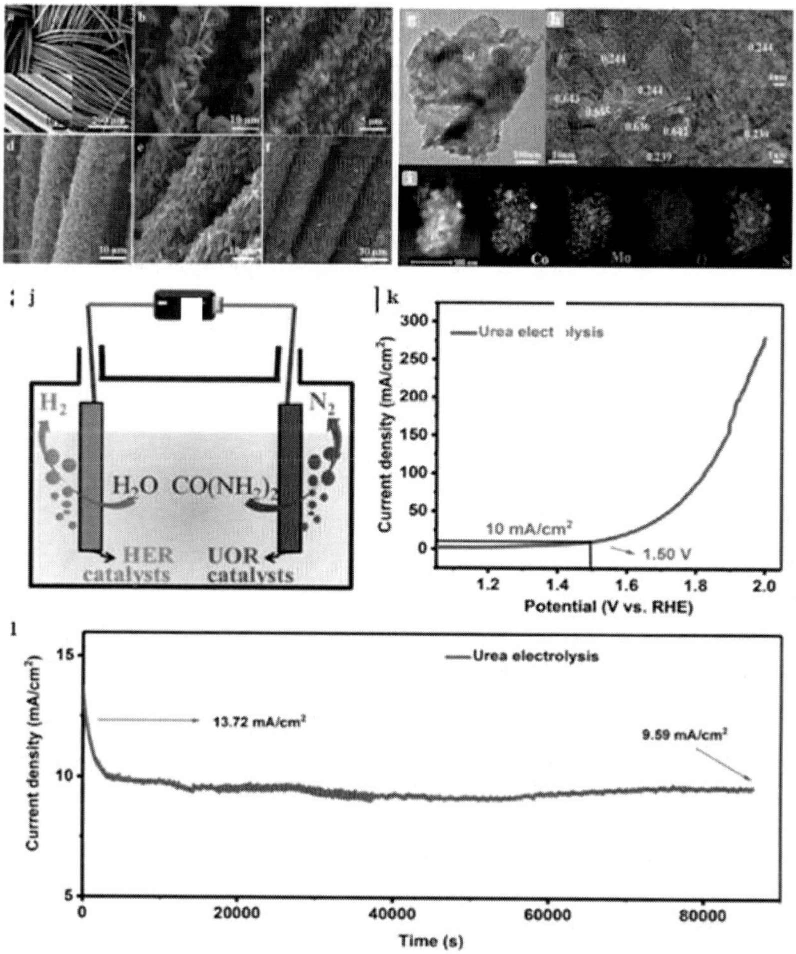

Figure 8. SEM images of (a) bare carbon cloth, (b) CoS_2-CC, (c) Co_5Mo_1S-CC, (d) Co_3Mo_1S-CC, (e) Co_1Mo_1S-CC, and (f) Co_1Mo_3S-CC. TEM (g) and HRTEM (h) images of Co_3Mo_1S-CC and the corresponding element mapping (i). (j) The schematic homemade urea electrolyzer using Co3Mo1S-CC as both the anode and cathode. (k) Polarization curves of Co3Mo1S-CC for urea electrolysis in 1 M KOH at a scan rate of 5 mV/s. (l) The long-term stability test of Co3Mo1S-CC towards urea electrolysis performed at a cell voltage of 1.52 V. Adapted with permission from Reference [67], Copyright (2020), American Chemical Society.

3.2. Ni- Free Bimetallic ECs

Till today, only a few of Ni-free catalysts have been synthesized which show good UOR performance when compared to Ni-based ECs. Even if the main catalyst comprises Ni-free composition, in most of the cases nickel foam (NF) has been used as a support for better dispersion and as conductive substrates. Gong et al. reported FeCo-LDH synthesized in situ on NF which showed excellent UOR activity with good durability. The EC required a voltage of 1.523 V and 1.409 V for water and urea electrolysis to obtain a current density that equals 10 mA cm^{-2}, respectively. The reason for such performance of the EC is due to the special morphology of the catalyst which was found to be crossed nanosheet structure with large surface area, facilitates rapid electron transfer [66].

Li and co-workers prepared a series of Co_xMo_yS nanosheets supported on carbon cloth (CC), out of which Co_3Mo_1S-CC showed bifunctional activity and durability for UOR and HER. The excellent activity of Co_3Mo_1S-CC is attributed to optimized structure and better electronic interaction between Mo and CoS_2 due to lower energy barrier. Also large available active reaction sites and good conductivity of the 3D self-supported structure enables the catalyst to meet the current density of 10 mA cm^{-2} at a very low overpotential of 85 mV [67]. Figure 8a illustrates SEM images of CC and Figure 8b-f illustrates surface of the CC is tightly covered with Co_xMo_yS nanosheets. When the ratio of Co/Mo is 3:1, then the surface of the carbon fibre is found to be of appreciable uniformity and present in a continuous fashion covered with nanosheets (Figure 8d), suggesting to be the optimum precursor ratio for the fabrication of Co_xMo_yS. Figure 8g, 8h shows the TEM and HRTEM images of Co_3Mo_1S-CC, respectively where formation of layered nanosheets can be confirmed. HRTEM image, further clearly indicates that the as synthesized Co_3Mo_1S is made up of crystalline CoS_2 and CoMoS structure. From the elemental mapping pattern (Figure 8h) uniform dispersion of Co, Mo and S is confirmed. Figure 8j is a scheme of Co_3Mo_1S-CC which can be used to fabricate both anode and cathode for a urea electrolyser. And only at an onset potential of 1.50 V, the EC achieves the current density of 10 mA cm^{-2} (Figure 8k) and the stability was done for 24 h where the current density dropped from 13.72 to 10 mA cm^{-2} and then remained around 9.59 mA cm^{-2} (Figure 8l). $Co_3V@C/Co_2VO_4$ (a carbon-encapsulated Co_3V decorated Co_2VO_4 nanosheets catalyst), was reported by Pan and co-workers which shows bifunctional activity. It required only a meagre potential of 1.31 V and −51 mV in 1.0 M KOH + 0.5 M urea solution to drive a current density of ±10 mA cm^{-2}. The excellent performance exhibited towards UOR and HER is attributed to the presence of hetero-junction between Co_3V and Co_2VO_4. Usually a hereto-junction catalyst can rearrange the electron in the interface to adjust and adapt to provide active sites and hence the synergy of the various active sites is used to further promote the kinetics of the reaction. Here, the hetero-junction between Co_3V and Co_2VO_4 provides active sites, and also the carbon encapsulation structure protects the catalyst from damage in strong alkaline solution. This altogether enhances the catalytic activity and stability of the EC [68].

Conclusion and Future Perspectives

The ORR and UOR are critical processes in electrochemical energy conversion and storage devices. The multiple-step, slow-kinetic e^- transfer in ORR and UOR constrains energy conversion efficiency. In-depth discussions of current research findings on the exquisite design of high-performance and cost-effective ORR and UOR ECs have been provided. There has been a significant advancement in the synthesis, characterization, and associated evaluation of ORR and UOR ECs. These newly established techniques, including alloy engineering, shape control, heteroatom doping, and surface strain engineering, have greatly improved the electrocatalytic performance of ORR and UOR ECs.

In conclusion, the application of bimetallic materials as ECs for applications in sustainable energy is still in its early stages, where opportunities and obstacles coexist. Future research initiatives can improve their performance by combining various strategies, such as developing ECs with a distinctive shape and morphology, introducing modified supports to increase surface area and conductivity, reducing agglomeration, incorporating other active species, and engineering their interface to speed up charge transport, etc. Despite these accomplishments, there is still much opportunity for thorough investigations, which can be focused on the following areas:

(1) Theoretical study leads to a better comprehension of ORR and UOR mechanisms. This study may make it easier to apprehend certain stages of the ORR and UOR pathway, such as adsorption, surface-bound intermediates, and transition state energetics. Additionally, a deeper understanding of the kinetics at the electrode-electrolyte

interface would enable the development of targeted approaches for efficiently designing ORR and UOR ECs.

(2) Identification of the ORR and UOR catalytic active sites. In general, there are two ways to increase the activity of an EC: (a) by raising the number of active sites on a specific electrode (for example, by nanostructure engineering an electrode to reveal more active sites) or (b) by boosting the intrinsic activity of each active site. Therefore, determining the type of active site is vital and ought to be the first step in designing an effective EC. An *in situ* study will also provide a fundamental understanding of the catalytic mechanism for both ORR and UOR. This is why developing in situ characterization techniques is essential, as they may reveal additional vital details about the intermediate and active sites and provide ideas for future EC design. The *in situ* characterization investigation will be possible using some methods, including near ambient pressure X-ray photoelectron spectroscopy, scanning probe microscopy, and soft X-ray absorption spectroscopy.

Likely, this field is poised for a transformative shift towards harnessing renewable energy through the utilization of cost-effective materials, driven by continuous research endeavors.

Acknowledgements

R. C. and S. P. B. thank Tezpur University (T. U.) for institutional fellowships. R. C. gratefully acknowledges T. U. for 'Research and Innovation Grant.'

References

[1] Chauhan, M., Reddy, K. P., Gopinath, C. S., Deka, S. Copper cobalt sulfide nanosheets realizing a promising electrocatalytic oxygen evolution reaction. *ACS Catal*. 2017, *7*(9), 5871-5879.

[2] Borah, B. J., Goswami, C., Yamada, Y., Tada, K., Tanaka, S., Bharali, P. Pd_2CuCo/C hybrid with nanoflower morphology toward oxygen reduction and formic acid oxidation reactions: Experimental and computational studies. *Energy & Fuels*. 2021, *35*(14), 11515-11524.

[3] Chen, M., Wang, L., Yang, H., Zhao, S., Xu, H., Wu, G. Nanocarbon/oxide composite catalysts for bifunctional oxygen reduction and evolution in reversible alkaline FCs: A mini review. *J. Power Sources*. 2018, *375*, 277-290.

[4] Osgood, H., Devaguptapu, S. V., Xu, H., Cho, J., Wu, G. Transition metal (Fe, Co, Ni, and Mn) oxides for oxygen reduction and evolution bifunctional catalysts in alkaline media. *Nano Today*. 2016, *11*(5), 601-625.

[5] Mekhilef, S., Saidur, R., Safari, A. Comparative study of different FC technologies. *Renewable Sustainable Energy Rev*. 2012, *16*(1), 981-989.

[6] Kirubakaran, A., Jain, S., Nema, R. K. A review on FC technologies and power electronic interface. *Renewable Sustainable Energy Rev*. 2009, *13*(9), 2430-2440.

[7] Carrette, L., Friedrich, K. A., Stimming, U. FCs: principles, types, fuels, and applications. *Chem Phys Chem*. 2000, *1*(4), 162-193.

[8] Ebrahimi, M., Kujawski, W., Fatyeyeva, K., Kujawa, J. A review on ionic liquids-based membranes for middle and high temperature polymer electrolyte membrane FCs (PEM FCs). *Int. J. Mol. Sci.* 2021, *22*(11), 5430.

[9] Modak, A., Mohan, R., Rajavelu, K., Cahan, R., Bendikov, T., Schechter, A. Metal–organic polymer-derived interconnected Fe–Ni alloy by carbon nanotubes as an advanced design of urea oxidation catalysts. *ACS Appl. Mater. Interfaces*. 2021, *13*(7), 8461-8473.

[10] Zhu, X., Dou, X., Dai, J., An, X., Guo, Y., Zhang, L., Tao, S., Zhao, J., Chu, W., Zeng, X. C., Wu, C. Metallic nickel hydroxide nanosheets give superior electrocatalytic oxidation of urea for FCs. *Angew. Chem. Int. Ed*. 2016, *55*(40), 12465-12469.

[11] Hazarika, K. K., Chutia, B., Changmai, R. R., Boruah, P. K., Das, M. R., Bharali, P. $Fe_xCo_{3-x}O_4$ nanohybrids anchored on a carbon matrix for high-performance oxygen electrocatalysis in alkaline media. *ChemElectroChem*. 2022, *9*(19), e202200867.

[12] Rai, V., Lee, K. P., Safanama, D., Adams, S., Blackwood, D. J. Oxygen reduction and evolution reaction (ORR and OER) bifunctional electrocatalyst operating in a wide pH range for cathodic application in Li–air batteries. *ACS Appl. Energy Mater*. 2020, *3*(9), 9417-9427.

[13] Jamesh, M. I. Recent progress on earth abundant hydrogen evolution reaction and oxygen evolution reaction bifunctional electrocatalyst for overall water splitting in alkaline media. *J. Power Sources*. 2016, *333*, 213-236.

[14] Periyasamy, S., Subramanian, P., Levi, E., Aurbach, D., Gedanken, A., Schechter, A. Exceptionally active and stable spinel nickel manganese oxide electrocatalysts for urea oxidation reaction. *ACS Appl. Mater. Interfaces*. 2016, *8*(19), 12176-12185.

[15] Goswami, C., Saikia, H., Tada, K., Tanaka, S., Sudarsanam, P., Bhargava, S. K., Bharali, P. Bimetallic palladium–nickel nanoparticles anchored on carbon as high-performance electrocatalysts for oxygen reduction and formic acid oxidation reactions. *ACS Appl. Energy Mater*. 2020, *3*(9), 9285-9295.

[16] Cui, X., Xiao, P., Wang, J., Zhou, M., Guo, W., Yang, Y., He, Y., Wang, Z., Yang, Y., Zhang, Y., Lin, Z. Highly branched metal alloy networks with superior activities for the methanol oxidation reaction. *Angew. Chem*. 2017, *129*(16), 4559-4564.

[17] Gutiérrez-Martín, F., García-De María, J. M., Baïri, A., Laraqi, N. Management strategies for surplus electricity loads using electrolytic hydrogen. *Int. J. Hydrog. Energy*. 2009, *34*(20), 8468-8475.

[18] Pareek, A., Dom, R., Gupta, J., Chandran, J, Adepu, V, Borse, P. H. Insights into renewable hydrogen energy: Recent advances and prospects. *Mater. Sci. Technol.* 2020, *3*, 319-327.

[19] Chen, L., Xu, X., Yang, W., Jia, J. Recent advances in carbon-based electrocatalysts for oxygen reduction reaction. *Chin Chem Lett*. 2020, *31*(3), 626-634.

[20] Zeng, K., Zheng, X., Li, C., Yan, J., Tian, J. H., Jin, C., Strasser, P., Yang, R. Recent advances in non-noble bifunctional oxygen electrocatalysts toward large-scale production. *Adv. Funct. Mater.* 2020, *30*(27), 2000503.

[21] Ma, Y., Ma, C., Wang, Y., Wang, K. Advanced nickel-based catalysts for urea oxidation reaction: challenges and developments. *Catalysts*. 2022, *12*(3), 337.

[22] Li, J., Wang, S., Chang, J., Feng, L. A review of Ni based powder catalyst for urea oxidation in assisting water splitting reaction. *Adv. Powder Technol.* 2022.

[23] Stacy, J., Regmi, Y. N., Leonard, B., Fan, M. The recent progress and future of oxygen reduction reaction catalysis: A review. *Renewable Sustainable Energy Rev.* 2017, *69*, 401-414.

[24] Xiong, D., Li, X., Fan, L., Bai, Z. Three-dimensional heteroatom-doped nanocarbon for metal-free oxygen reduction electrocatalysis: A review. *Catalysts* 2018, *8*(8), 301.

[25] Shao, M., Chang, Q., Dodelet, J. P., Chenitz, R. Recent advances in electrocatalysts for oxygen reduction reaction. *Chem. Rev.* 2016, *116*(6), 3594-3657.

[26] Porter, N. S., Wu, H., Quan, Z., Fang, J. Shape-control and electrocatalytic activity-enhancement of Pt-based bimetallic nanocrystals. *Acc. Chem. Res.* 2013, *46*(8), 1867-1877.

[27] Zhou, M., Wang, H. L., Guo, S. Towards high-efficiency nanoelectrocatalysts for oxygen reduction through engineering advanced carbon nanomaterials. *Chem. Soc. Rev.* 2016, *45*(5), 1273-1307.

[28] Nørskov, J. K., Rossmeisl, J., Logadottir, A., Lindqvist, L. R. K. J., Kitchin, J. R., Bligaard, T., Jonsson, H. Origin of the overpotential for oxygen reduction at a fuel-cell cathode. *J. Physic. Chem. B* 2004, *108*(46), 17886-17892.

[29] Sanij, F. D., Balakrishnan, P., Leung, P., Shah, A., Su, H., Xu, Q. Advanced Pd-based nanomaterials for electro-catalytic oxygen reduction in fuel cells: A review. *Int. J. Hydrog. Energy* 2021, *46*(27), 14596-14627.

[30] Lin, R., Zheng, T., Chen, L., Wang, H., Cai, X., Sun, Y., Hao, Z. Anchored Pt-Co nanoparticles on honeycombed graphene as highly durable catalysts for the oxygen reduction reaction. *ACS Appl. Mater. Interfaces* 2021, *13*(29), 34397-34409.

[31] Kim, H. Y., Kwon, T., Ha, Y., Jun, M., Baik, H., Jeong, H. Y., Kim, H., Lee, K., Joo, S. H. Intermetallic PtCu nanoframes as efficient oxygen reduction electrocatalysts. *Nano Lett.* 2020, *20*(10), 7413-7421.

[32] Mao, L., Fu, K., Jin, J., Yang, S., Li, G. PtFe alloy catalyst supported on porous carbon nanofiber with high activity and durability for oxygen reduction reaction. *Int. J. Hydrog. Energy* 2019, *44*(33), 18083-18092.

[33] Cai, X., Lin, R., Shen, D., Zhu, Y. Gram-scale synthesis of well-dispersed shape-controlled Pt– Ni/C as high-performance catalysts for the oxygen reduction reaction. *ACS Appl. Mater. Interfaces* 2019, *11*(33), 29689-29697.

[34] Guo, Y. Z., Yan, S. Y., Liu, C. W., Chou, T. F., Wang, J. H., Wang, K. W. The enhanced oxygen reduction reaction performance on PtSn nanowires: the importance of segregation energy and morphological effects. *J. Mater. Chem. A* 2017, *5*(27), 14355-14364.

[35] Lai, Y., Du, G., Zheng, Z., Dong, Y., Li, H., Kuang, Q., Xie, Z. Facile synthesis of clean PtAg dendritic nanostructures with enhanced electrochemical properties. *Inorg. Chem. Front.* 2020, *7*(5), 1250-1256.

[36] Goswami, C., Saikia, H., Tada, K., Tanaka, S., Sudarsanam, P., Bhargava, S. K., Bharali, P. Bimetallic palladium–nickel nanoparticles anchored on carbon as high-performance electrocatalysts for oxygen reduction and formic acid oxidation reactions. *ACS Appl. Energy Mater.* 2020, *3*(9), 9285-9295.

[37] Zamora Zeledón, J. A., Stevens, M. B., Gunasooriya, G. T., Gallo, A., Landers, A.T., Kreider, M. E., Hahn, C., Nørskov, J. K., Jaramillo, T. F. Tuning the electronic structure of Ag-Pd alloys to enhance performance for alkaline oxygen reduction. *Nat. Commun.* 2021, *12*(1), 1-9.

[38] Zhang, Z., Liu, S., Tian, X., Wang, J., Xu, P., Xiao, F., Wang, S. Facile synthesis of N-doped porous carbon encapsulated bimetallic PdCo as a highly active and durable electrocatalyst for oxygen reduction and ethanol oxidation. *J. Mater. Chem. A* 2017, *5*(22), 10876-10884.

[39] Wang, Y., Zhu, Z., Xu, K., Guo, W., Yu, T., He, M., Wei, W., Yang, T. Palladium nanobelts with expanded lattice spacing for electrochemical oxygen reduction in alkaline media. *ACS Appl. Nano Mater.* 2021, *4*(2), 2118-2125.

[40] Zhang, L. Y., Guo, C. X., Cao, H., Wang, S., Ouyang, Y., Xu, B., Guo, P., Li, C. M. Highly wrinkled palladium nanosheets as advanced electrocatalysts for the oxygen reduction reaction in acidic medium. *Chem. Eng. J.* 2022, *431*, 133237.

[41] Liu, M., Lyu, Z., Zhang, Y., Chen, R., Xie, M., Xia, Y. Twin-directed deposition of Pt on Pd icosahedral nanocrystals for catalysts with enhanced activity and durability toward oxygen reduction. *Nano Lett.* 2021, *21*(5), 2248-2254.

[42] Xia, W., Mahmood, A., Liang, Z., Zou, R., Guo, S. Earth-abundant nanomaterials for oxygen reduction. *Angew. Chem. Int. Ed.* 2016, *55*(8), 2650-2676.

[43] Huang, Z. F., Wang, J., Peng, Y., Jung, C. Y., Fisher, A., Wang, X. Design of efficient bifunctional oxygen reduction/evolution electrocatalyst: recent advances and perspectives. *Adv. Energy Mater.* 2017, *7*(23), 1700544.

[44] Singh, S. K., Kashyap, V., Manna, N., Bhange, S. N., Soni, R., Boukherroub, R., Szunerits, S., Kurungot, S. Efficient and durable oxygen reduction electrocatalyst based on CoMn alloy oxide nanoparticles supported over N-doped porous graphene. *ACS Catal.* 2017, *7*(10), 6700-6710.

[45] Lu, Y., Zhang, N., An, L., Li, X., Xia, D. Synthesis of high dispersed intermetallic Ag_4Sn/C and its enhanced oxygen reduction reaction activity. *J. Power Sources* 2013, *240*, 606-611.

[46] Nguyen, D. M., Bach, L. G., Bui, Q. B. A high-performance catalyst based on binary CuFe alloyed nanocrystals encapsulated in nitrogen-doped graphene nanosheets towards oxygen reduction reaction. *J. Solid State Chem.* 2019, *273*, 132-140.

[47] Wang, J., Chen, F., Jin, Y., Lei, Y. Dilute Au-containing Ag nanosponges as a highly active and durable electrocatalyst for oxygen reduction and alcohol oxidation reactions. *ACS Appl. Mater. Interfaces* 2018, *10*(7), 6276-6287.

[48] Wang, J., Huang, Z., Liu, W., Chang, C., Tang, H., Li, Z., Chen, W., Jia, C., Yao, T., Wei, S., Wu, Y. Design of N-coordinated dual-metal sites: a stable and active Pt-free catalyst for acidic oxygen reduction reaction. *J. Am.* Chem. Soc. 2017, 139(48), 17281-17284.

[49] Hu, H., Liang, J. H., Zu, Z. Y., Mi, J. L., Xiao, B. B., Zhang, P. Ni/Cu regulating nitrogen-doped porous carbon as electrocatalyst for oxygen reduction reaction. *ChemistrySelect* 2021, *6*(27), 6949-6956.

[50] Zhu, B., Liang, Z., Zou, R. Designing advanced catalysts for energy conversion based on urea oxidation reaction. *Small*. 2020, *16*(7), 1906133.

[51] Isaka, Y., Kato, S., Hong, D., Suenobu, T., Yamada, Y., Fukuzumi, S. Bottom-up and top-down methods to improve catalytic reactivity for photocatalytic production of hydrogen peroxide using a Ru-complex and water oxidation catalysts. *J. Mater. Chem*. 2015, *3*(23), 12404-12412.

[52] Xu, Y., Ren, T., Ren, K., Yu, S., Liu, M., Wang, Z., Li, X., Wang, L., Wang, H. Metal-organic frameworks-derived Ru-doped Co_2P/N-doped carbon composite nanosheet arrays as bifunctional electrocatalysts for hydrogen evolution and urea oxidation. *J. Chem. Eng*. 2021, *408*, 127308.

[53] Wang, G., Ling, Y., Lu, X., Wang, H., Qian, F., Tong, Y., Li, Y. Solar driven hydrogen releasing from urea and human urine. *Energy Environ. Sci*. 2012, *5*(8), 8215-8219.

[54] Li, C., Liu, Y., Zhuo, Z., Ju, H., Li, D., Guo, Y., Wu, X., Li, H., Zhai, T. Local charge distribution engineered by schottky heterojunctions toward urea electrolysis. *Adv. Energy Mater.* 2018, *8*(27), 1801775.

[55] Xu, W., Wu, Z., Tao, S. Urea-based FCs and electrocatalysts for urea oxidation. *Energy Technol.* 2016, *4*(11), 1329-1337.

[56] Bolzan, A. E., Iwasita, T. Determination of the volatile products during urea oxidation on platinum by on line mass spectroscopy. *Electrochim. Acta*. 1988, *33*(1), 109-112.

[57] Boggs, B. K., King, R. L., Botte, G. G. Urea electrolysis: direct hydrogen production from urine. *ChemComm*. 2009, *32*, 4859-4861.

[58] Hu, X., Zhu, J., Li, J., Wu, Q. Urea electrooxidation: current development and understanding of Ni-based catalysts. *ChemElectroChem*. 2020, *7*(15), 3211-3228.

[59] Vedharathinam, V., Botte, G. G. Understanding the electro-catalytic oxidation mechanism of urea on nickel electrodes in alkaline medium. *Electrochim. Acta.* 2012, *81*, 292-300.

[60] Vedharathinam, V., Botte, G. G. Direct evidence of the mechanism for the electro-oxidation of urea on $Ni(OH)_2$ catalyst in alkaline medium. *Electrochim. Acta*. 2013, *108*, 660-665.

[61] Daramola, D. A., Singh, D., Botte, G. G. Dissociation rates of urea in the presence of NiOOH catalyst: a DFT analysis. *J. Phys. Chem.* 2010, *114*(43), 11513-11521.

[62] Munde, A. V., Mulik, B. B., Dighole, R. P., Sathe, B. R. Urea electro-oxidation catalyzed by an efficient and highly stable Ni–Bi bimetallic nanoparticles. *ACS Appl. Energy Mater.* 2021, *4*(11), 13172-13182.

[63] Yang, X., Zhang, H., Xu, W., Yu, B., Liu, Y., Wu, Z. A doping element improving the properties of catalysis: in situ Raman spectroscopy insights into Mn-doped NiMn layered double hydroxide for the urea oxidation reaction. *Catal. Sci. Technol.* 2022, *12*(14), 4471-4485.

[64] King, R. L., Botte, G. G. Investigation of multi-metal catalysts for stable hydrogen production via urea electrolysis. *J. Power Sources.* 2011, *196*(22), 9579-9584.

[65] Li, Y., Jiang, H., Cui, Z., Zhu, S., Li, Z., Wu, S., Ma, L, Han, X., Liang Y. Spin state tuning of the octahedral sites in Ni–Co-based spinel toward highly efficient urea oxidation reaction. *J. Phys. Chem. C.* 2021, *125*(17), 9190-9199.

[66] Gong, Y., Zhao, H., Ye, D., Duan, H., Tang, Y., He, T., Shah, L. A., Zhang, J. High efficiency UOR electrocatalyst based on crossed nanosheet structured FeCo-LDH for hydrogen production. *Appl. Catal. A: Gen.* 2022, *643*, 118745.

[67] Li, P., Zhuang. Z., Du, C., Xiang, D., Zheng, F., Zhang, Z., Fang, Z., Guo, J. Zhu, S., Chen, W. Insights into the Mo-doping effect on the electrocatalytic performance of hierarchical Co_xMo_yS nanosheet arrays for hydrogen generation and urea oxidation. *ACS Appl. Mater. Interfaces.* 2020, *12*(36), 40194-40203.

[68] Pan, M., Chen, W., Qian, G., Yu, T., Wang, Z., Luo, L., Yin, S. Carbon-encapsulated Co_3V decorated Co_2VO_4 nanosheets for enhanced urea oxidation and hydrogen evolution reaction. *Electrochim. Acta.* 2022, *407*, 139882.

Chapter 4

Bimetallic Nano-Systems and Their Applications as Sensors

Jejiron Maheswari Baruah[1]
Samina Afsara Hussain[2]
Jayanta Kumar Sarmah[2,3]
and Satyabrat Gogoi[2,*]
[1]Department of Chemistry, North Lakhimpur College (Autonomous), North Lakhimpur, Assam, India
[2]The Assam Kaziranga University, Department of Chemistry, School of Basic Sciences, Jorhat, Assam, India
[3]Department of Chemistry, Rabindranath Tagore University, Hojai, Assam, India

1. Introduction

Since their inceptions, nanomaterials and nano-based techniques have gained tremendous interest in different domains. It is a well-established fact that the nanoscale dimension can bring unique properties to materials which otherwise is not possible to achieve in bulk state. Researchers are constantly involving for value addition to nanomaterials in terms of chemical, physical, electrical, optical, biological, catalytic properties that can be used to solve real life problems. In this context, metallic nanoparticles have been explored extensively in the past few decades. They exhibit unique properties including superior electrical, catalytic, and optical behavior. Metallic nanomaterials can be obtained in the noble form as well as in the form of oxides and sulfides [1]. Consistent efforts are going on to improve various properties of metallic

* Corresponding Author's Email: gogoisb@gmail.com.

In: Bimetals
Editors: Pranjal Saikia, Jayanta Kumar Sarmah and Parikshit GogoiName
ISBN: 979-8-89113-496-6

nanoparticles. One such approach is to constitute bimetallic nano-system. A bimetallic nano-system typically consist of two metal-based nanomaterials of different origin. The idea of bimetallic nanomaterials is to enhance the native properties of monometallic system through synergic interaction of both the component [2]. In many instances newer attributes can also be imparted to a bimetallic system. Controlled structure and composition are the two important parameters that can add tailor made properties to a bimetallic system. Contemporary literature reveals a large number of works on bimetal synthesis, characterization, property exploration and applications.

Among various applications of bimetallic nanosystem, one important one is their utilization in sensing platforms [3-4]. A sensor is a platform that can detect and measure analyte by creating signals with analyte concentration. A sensor typically consists of a recognition element, transducer, amplifier and detector [4]. Recognition element engages the target analyte specifically with the probe. The function of the transducer is to create readable signals based on the interaction of recognition element and the target analyte. The amplifier then enhanced the signal strength which displayed after analyzing by the detector. Based on the types of signals different transducer can be classified into different categories like colorimetric, electrochemical, optical etc. Since metals, including bimetallic systems, possesses superior catalytic, optical, electrical and biological properties, their inclusion in sensors creates enormous opportunities towards development of highly sensitive and specific detection of various analytes. Due to their unique catalytic activity, they can be used as nano-enzyme in many cases, where they act as analyte recognition unit. The catalytic action of a bimetallic system can generate signals against specific analytes in terms of voltage, current, absorption, or emission. The bimetallic systems can play a multifunctional role in the development of sensing platforms. Therefore, this chapter intends to make a concise description on bimetallic systems, their properties and application in terms of sensing platforms based on contemporary art of literature.

2. Different Types of Bimetallic Compounds: Structure, Property and Application

Bimetallic nanocrystals are comprised of two elements only, their architecture can be expansively coordinated with respect to configuration (core-shell, dendritic or dimeric), alloyed or intermetallic atomic ordering, crystal structure (with various stacking faults or twin defects) (Figure 1). However, this variation may be confided both upon thermodynamic and kinetic stability of these nanostructures. In this section, we are confining this divergence with respect to the spatial distribution of the nanocrystals.

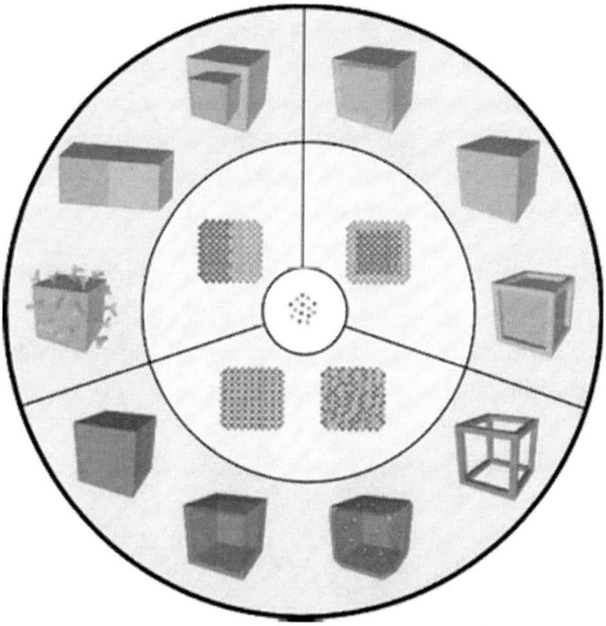

Figure 1. Schematic diagram of bimetallic nanocrystals with distinct architectures (outer ring), the evolution of atomic distributions from two types of metal atoms (center) to four types (middle ring). [5].

2.1. Core–Shell Nanocrystals

A core-shell nanocrystal, denoted as core@shell, consists of a sharply defined inner core surrounded by a shell made of a different metal [6].

Synthesis

The seed-mediated approach utilising either the thermal decomposition or chemical reduction is primarily adopted for the synthesis of core-shell material [7-10]. This method was originally developed using colloidal particle templates (e.g., silica spheres, metal-oxide particles, etc.) and is adaptable to various geometries, including wires, spheres, prisms, etc., [6]. In contemporary research, this synthesis technique has advanced significance which enables the production of core-shell nanocrystals with precise shell thicknesses, even up to one atomic layer.

Applications

The tunability of shell thickness can significantly impact several applications, which can be categorized as (i) enhanced stability in thermal and chemical domains, (ii) improved catalytic activity through strain effect or electronic coupling, (iii) an increase in magnetic anisotropy, (iv) tuning of optical properties, and (v) provision to act as a critical intermediate in the formation of structures such as cages, boxes etc.

Classification of Core-Shell Material

a. *Shell thickness:* The thickness of the shell may influence the physicochemical properties tremendously. Hence depending on the shell thickness core-shell nanocrystals are further categorized into three different subgroups: (a) a single atomic layer, (b) 2-6 atomic layers and (c) more than 6 atomic layers.
b. *Variation in core-shell pattern:* Moreover, depending upon the type of core-shell, they can further be classified as shown in Figure 2.

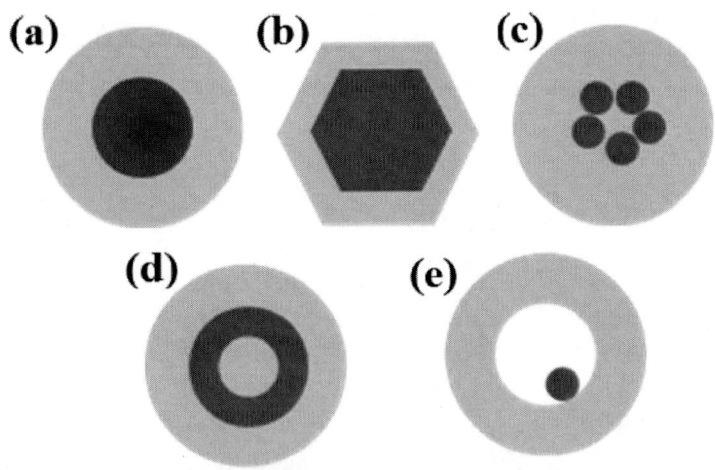

Figure 2. Different core/shell nanoparticles: (a) spherical core/shell nanoparticles; (b) hexagonal core/shell nanoparticles; (c) multiple small core materials coated by single shell material; (d) nanomatryushka material; (e) movable core within hollow shell material [6].

2.1.1. Shell Thickness

(a) Single Atomic Layer (Skin Layer) Thick Shell

In this type of core-shell, a single atomic layer ingrains a nanocrystal core. Such nanocrystal is denoted by M@N1L. This can be used in heterogeneous catalysis in which atoms upon surface, having lowest cost materials, is get emerged with their immense probable utilization efficacy. The specific activity of the catalytic system is get maximized due to heteroepitaxial strains imposed by the core upon shell material together with the electronic coupling.

There are processes which can be adopted for controllably growing a monolayer and can be enlisted as: (i) formation of a sacrificial monolayer (e.g., Cu) via under-potential deposition (UPD) and galvanic replacement (ii) seed-mediated growth and (iii) formation of an alloy or intermetallic nanocrystal, followed by inducing phase segregation.

(b) Shells of Multiple (n = 2-6) Atomic Layers Thick

Another class of core-shell materials in which the core is encapsulated by 2-6 atomic layers, denoted by M@NnL (n=number of atomic layers) and its thickness can offer electronic coupling and strain between the shell and the core. Since the shell is having a thickness of more than six atomic layers, hence they are considered to be free from property modulation induced by the core material. Moreover, it seems they depend only upon of the shell metal. Therefore, in this subclass of core-shell material, the thickness-dependent catalytic activity is at its most advantageous position. For example, in a recent study of Pd@PtnL icosahedra (n = 0.7, 2.0, 2.7, 4.8), the optimal shell thickness for achieving the highest ORR mass activity was determined within the intermediate thickness of n = 2.7. [11], which eventually highlights capability of depositing material with monolayer accuracy.

Synthesis

Four methods are generally adopted for controllably generating M@NnL (n = 2-6) nanocrystals:

(i) seed-mediated growth,
(ii) sequentially repeating the UPD and galvanic replacement processes,
(iii) phase segregation [12] and
(iv) anion coordination [13].

(c) Shells Thicker Than Six Atomic Layers

Shells more than six atomic layers (or 1.5 nm) thick has established themselves in majority of research literature including both the synthesis and application of core-shell nanoparticles. Here due to the diminishing electronic coupling between the outermost layer of the shell and core, the catalytic activity depending upon strain has vanished. However, the enlisted advantages of this type of core-shell nanocrystals are:

(i) Broad range tuning of optical properties,
(ii) improvised chemical and thermal stability and
(iii) enhanced magnetic properties.

Though the shell thickness is sensitive towards LSPR peak intensity and position, still the optical response is get dominated by the shell metal's critical thickness, pretended as if the shell shields the core. Hence, tailoring the thickness of the shell is of utmost necessity to enhance the optical properties of core-shell nanocrystals.

Synthesis

The synthetic approach is most commonly the seed-mediated growth. There is another type of core-shell structure labelled as *Hollow derivatives of core-shell nanocrystals*, exemplified by nanocages and nanoboxes having the capability of tuning their optoelectronic capability over the entire range of visible and near-infrared (NIR) [14-15]. Though both nanoboxes and nanocages belong to the same broad category, but their porous walls draw the line of difference. As pores in Nanocages allow easy penetration of small molecules, solvent etc., through both backward and forward movement, they are exemplary for drug delivery, encapsulation, controlled release and catalytic activity than their solid correspondent [16-18]. On the contrary, pore-free walls of Nano boxes can efficaciously quarantine the interiors from the external environment.

Synthesis

(i) In the core-shell nanocrystals, selective etching away the core from it to leave a cage behind,
(ii) galvanic replacement,
(iii) using the *Kirkendall effect*,
(iv) combination of all the cited approaches.

Application of Hollow Derivatives of Core-Shell Nanocrystals:
Such optical tunability due to highly open structure can enable advanced applications in biomedicine *viz* photothermal cancer therapy, sensing, drug-delivery etc. [16, 19].

2.2. Dendritic Nanocrystals

The elementary idea of Dendritic is branching. Thus, the structural type of bimetallic nanocrystals, which represents multipods, pentacles, highly branched nanocrystals, sea urchin-like, stars, flowers etc., as well as heterostructured forms (*viz* dumbbell structures, Janus particles, dimers etc.) are belonged to the Dendritic and heterostructured class of nanocrystals. [20-29].

Synthesis
Four general routes to synthesize dendritic nanocrystals are: (i) galvanic replacement, (ii) seed-mediated growth (kinetically controlled), iii) aggregation or oriented attachment and (iv) anisotropic etching of facet-controlled nanocrystals. A heterostructured bimetallic nanocrystal can be generated by joining a few interfaces of two metals. They can be classified on the basis of symmetry breaking, where seed-mediated process (kinetically controlled) is considered for their genesis. Their formation is favoured when non-epitaxial deposition can be made or when the rate of surface diffusion is retarded by adopting low temperature.

Application

a. They are important in heterogeneous catalysis appeal due to their accessible high index surfaces and large surface-to-volume ratios (peculiar high curvature around the branches as well as at tips).
b. The stinging tips are also liable for concentrating electrons which generates profound electric fields which are significant for optical applications *viz* Surface Enhanced Raman Scattering (SERS).
c. Sensing application can also be made due to their high sensitization towards minute changes in local refractive index (e.g., exchange of ligand, change of solvent, binding events, etc.) occurs by virtue of highly branched plasmonic structures.

2.3. Atomic Ordering

Intermetallic compounds and *Alloys* are two basic distinctions of Atomic Ordering in bimetallic nanocrystals. Basically, an intermetallic compound is attributed to a system having a well-defined stoichiometry with long-range atomic order. However, alloy can be introduced as nanocrystal building block of two metals associated randomly, however this is to be done thoroughly. This specified divergence in structure of these two major classes provides a clear picture about their different properties with indistinguishable atomic ratio and elemental composition.

Though thermodynamic stability (*viz* lattice parameters, atomic interactions etc.) provides uncloudy information about the most encouraging position of atoms both in intermetallic and facet, however, kinetics also plays an important role in this theme [30-33].

2.3.1. Alloys

In this form of bimetallic nanocrystals, research reveals that they are superior in the manifestation of unique physicochemical properties than individual nanocrystals. [34-36].

Synthesis

a. Co-reduction,
b. Thermal decomposition- difficult to control both composition and shape simultaneously,
c. Galvanic replacement,
d. Kirkendall effect- for hollow nanocrystal production.

The affirmative mixing of two metals can occur when,

a. The dissimilar metals have stronger bond than both the similar metal-metal bonds;
b. The two metals have similar lattice parameters *viz* lattice constant etc.;
c. The surface of the two metals have comparable energies [37].

Nevertheless, these parameters portray a few thermodynamically stable alloys, but possibility of forming a non-equilibrium phase during the synthesis is also there. At relatively low temperature, the reaction prevents the atoms from grasping their thermodynamically recommended positions through bulk and surface diffusion. In such cases, possibility of formation of dimeric structure or core-shell has become more common than to observe miscibility of metals, i.e., alloy creation [20]. Hence, heating is a compulsion to stimulate the diffusion among the atoms to boost the generation of alloy.

2.3.2. Intermetallic Compounds

When well-defined stoichiometry precedents over the formation of bimetallic nanocrystals than random mixing or alloying, then this type of construction is termed as *Intermetallic Compounds*. Thermodynamically, prediction can be made towards the dominance in the formation of an alloy or intermetallic compound by considering the change in Gibbs free-energy upon mixing (ΔG_{mix}): [38]

$$\Delta G_{mix} = \Delta H_{mix} - T\Delta S_{mix} \tag{1}$$

where,

ΔH_{mix} is change in enthalpy and ΔS_{mix} is the change in entropy during mixing

In principle, the atomic ordering (under a given experimental conditions) in a bimetallic system is determined by minimization of ΔG_{mix}. It is also important that *equation 1* is not considering other energetic terms *viz* surface-free energies, defect-free energies and strain-free energies, which plays an important role towards the formation of desired nanocrystals. Hence, in most cases, intermetallic compounds end up in alloys with disordered atomic arrangements and in order to acquire an intermetallic compound from alloy, post-synthesis annealing (under inert condition) has become mandatory [11].

Application

a. Intermetallic nanomaterials have demonstrated their supremacy in the domain of magnetism and catalysis [39-40].
b. In the context of oxygen reduction reaction (ORR) via *electrocatalysis*, Pt–Ni-based intermetallic nanocrystals have displayed magnificent activity.
c. In future information storage programme, magnetism exhibited by intermetallic compounds have been also identified.

3. Bimetal as Sensors

Bimetallic nano-systems can act as sensing platforms due to their unique catalytic, electrical and optical properties. Compared to monometallic systems, bimetallic systems have many advantages for sensor applications. Bimetallic nano-stems possess enhanced sensitivity because of synergistic effect of the two metallic constituents. Such systems also possess high stability as bimetallic systems are due to metallization provides better resistance towards oxidation [4]. Bimetallic sensors claim to possess high sensitivity, which is due to enhanced bio-mimetic activity of muti-metal systems. They also possess superior biocompatibility compared to monometallic system. Due to the presence of different metal systems, biocompatibility of bimetallic sensors can be tailormade which makes them less toxic. Based on the characteristics of the nano-system different types of detection mechanism can be followed. That includes colorimetry-based detection, electro-chemical detection, fluorescence detection, chemiluminescence etc. [1]. Colorimetric detection relies on the catalytic activity of bimetal systems. One of the classic examples is peroxidase like catalytic activity of metal-based nanoparticles. Such nano systems can specifically catalyse oxidation of chemicals like TMB [41]. TMB is a colourless compound and exhibits blue colour in oxidized form. Hence, any substance that can interfere the oxidation of TMB both in positive and negative ways can be detected by following colorimetry-based methods. On the other hand, electrochemical sensors rely on ionic conductivity of the electrodes and electrochemically active species which are generated during electrochemical reactions [1]. Metals are the most popular materials for the construction and modification of electrodes. Due to enhanced electronic property and catalytic activity, bimetallic systems provide better ionic conductivity which leads to sensitive detection of many different analytes. Fluorescence based detection entirely depend on optical properties of bimetallic systems. It is a known fact that many metals in nano state possess unique fluorescence property. Such property can be further enhanced by constituting bimetallic system. In certain bimetal systems dual emission can also achieved. Such bimetals can be used to constitute ratiometric based fluorescence sensors. In such case, fluorescence intensity of one emission is affected by the presence of analyte to be detected, whereas the second emission is not affected. Such relative change in intensity provides highly sensitive detection which can also be detected by observing change in luminescence of colour under a suitable excitation source. Here, we are trying to discuss some important chemical entities that can be detected by bimetal systems.

Glucose is one molecule which has been detected by bimetal-based sensors successfully. Glucose can be detected by following two different approaches, *viz*, enzymatic detection and non-enzymatic detection [4]. Bimetal based systems exhibit peroxidase mimic activity. Therefore, they are suitable to construct colorimetry-based sensing platforms. It has been a known fact that metallic nanoparticles possess peroxidase like activity. Such activity can be further enhanced with the help of bimetallic system. Bimetallic system of Au and Ag are widely reported for colorimetry detection of glucose and hydrogen peroxide. One common approach is to use TMB, which is oxidised by Au/Ag system into ox-TMB. In the oxidized form, TMB has blue colour, but in presence of glucose the colour disappears in a concentration wise manner which is analytically followed by recording the absorption. The relationship between change in absorption intensity and concentration gives the amount of glucose present [42]. On the other hand, quantification of H_2O_2 is rather direct. Quantitative estimation of ox-TMB by Spectro-photochemically directly gives the amount of H_2O_2 present. Glucose can also be detected by constructing electrochemical based bimetallic systems. Both enzymatic and non-enzymatic methods have been developed in recent time. In non-enzymatic approach Pt based electrodes can be used effectively. Pt alone as a glucose electrode possess certain limitations; such as slow glucose oxidation, adsorption of chloride on electrode surface and interference from other biological entities. Therefore, inclusion of bimetallic system with Pt as a component can enhance efficiency of the detection process. Different metallic systems such as Ir, Bi, Ni, Pd, Au, Ru, Pb, and Rh have been reported to possess good electrochemical behaviour. Pt-Au, Pt-Ru, Pt-Ir, Pd/Au, Au/Ru have been reported to be used as bimetallic glucose electrode. Noble gas/transition metal and transition metal/transition metal-based systems have also been reported with excellent electrochemical properties. Examples includes Pt/Pb, Pt/Cu, Au/Cu based electrodes. Among transition metal based bimetallic sensors, Ni and Cu are important [43]. In is pertinent to mention that in recent time activated carbon-based nanomaterials are also included as support for bimetallic systems due to the superior electrical properties of carbonaceous nanostructures. MWCNT and graphene-based nanostructures are gaining tremendous interest in recent time. During glucose oxidation H2O2 is generated at anodic potential, which is then detected by electrochemically. GOx is immobilized with bimetallic nano system serves as the electrode in the whole detection process. Bimetal due to unique architecture and composition provides better electric properties which enhanced the sensing performance towards glucose.

In case of cholesterol, utilization of bimetallic system is relatively rare. However, Pt-Pd system has also been reported to use for cholesterol detection through an enzymatic approach and by adopting electrochemical procedure. The electrode for cholesterol was built with the help of GCE modified with Pt-Pd doped graphene and chitosan. The nanocomposite was linked with cholesterol oxidase (COX) enzyme to bring high specificity and sensitivity [44]. Ag coated Au core-shell was used for colorimetric detection of cholesterol. Cholesterol oxidase (COx) was used as enzyme recognition element, which generates H2O2 quantitatively by reacting with cholesterol. H2O2 thus produced causes etching of Ag from Au core, which results change in colour and change in absorption maximum [42]. Bimetallic Eu/Fe-MOFs has been reported for the detection of cholesterol by using ratiometric fluorescence method. This bimetallic system possesses dual emission at 458 nm and 615 nm which corresponded to MIL-53(Fe) and Eu centres respectively. However, in presence of cholesterol Eu emission get enhanced while, Fe emission gets quenched. This causes a relative change in fluorescence intensity with the amount of cholesterol that form the basis of cholesterol detection [45].

Protein and nucleic acids can also be detected by using bimetallic systems. Ag/Zn bimetallic metal-organic framework is reported for label free detection of unamplified HCV nucleic acid. Synergistic catalytic effect of Ag and Zn causes signal amplification which facilitates highly sensitive detection of target nucleic acid [46]. Similarly, Co-Ni-based metal–organic framework was used in an electrochemical based sensor for detection of miRNA-126. Co-Ni-based metal–organic framework exhibits high electrochemical activity because of synergic effect of multiple metal centres. With this bimetal system complementary DNA of miRNA-126 was immobilized. This ensures a very specific interaction of miRNA-126 over the bimetal system. Such interaction causes an amplified electrochemical response which is used to detect miRNA-126 at an ultrasensitive way [47]. Even multi-metal system can also be used to develop new sensors. In such an attempt Au-Ag-Pt based system was used for Real-time nucleic acid detection via field-effect transistor sensors (FET). The probe comprises of two Au-electrodes connected by a graphite oxide channel. The graphite oxide channel was decorated with try-nanocluster system that contains Au-Ag-Pt. With increase in concentration of DNA can amplified the current signal from which nucleic acid is detected effectively [48].

Different microorganism can also be detected with the help of bimetal-based sensors. For example, *Staphylococcus aureus* bacteria in food samples can be detected by a paper based colorimetric method that uses Au/Pt bimetallic nanoclusters. The reported Au/Pt bimetallic system based on oligonucleotide base possess very high peroxidase like activity. It can oxidize TMB which has deep blue colour and generate H2O2 during the reaction. However, in presence of *Staphylococcus aureus,* oxidation of TMB gets prohibited that causes decrease in colour intensity of oxidized-TMB. The whole colour change can be followed with naked eye and photoelectrochemically as well [49]. Ag/Au bimetallic nanoparticles has been reported for label free discrimination of three bacterial stains viz. *Escherichia coli, Salmonella typhimurium,* and *Bacillus subtilis.* The whole detection relies on a phenomenon called surface-enhanced Raman scattering (SERS). The approach relies on a priori synthesis of Ag/Au bmNPs and subsequent SERS studies on bacteria [50]. Bimetallic Au-Pt was reported to use in an aptamer-based detection of *Escherichia coli O157:H7*. The synergistic effect of Au and Pt imparts high enzyme like catalytic activity in the nano system. Upon irradiated by light source, the Au-Pt nano-catalyst system induces high photochemical conversion efficiency. Such characteristics is applied for the eradication and detection of *E. coli O157:H7* with an extreme efficacy (95%) and within a very quick time of 5 min [51].

In addition to above mentioned biomolecules and bio-entities, different fine chemicals can also be detected by using bimetal systems. Organophosphorus based pesticides are well detectable by bimetal sensors. Different combinations of metallic systems are known. Bimetallic Pd-Au system has been reported to use for the acetylcholinesterase-based electrochemical detection of organophosphate pesticides. Acetylcholinesterase was immobilized over Pd-Au nanosystem to bring high sensitivity of the sensor towards organophosphate pesticides. The bimetallic system was used to modified a glassy carbon electrode which served as the sensor probe. The amperometric response was used for the quantitative estimation of organophosphate pesticides [52]. Similarly, fluorescence-based methods can also be adopted. A bimetallic Ag-Au has been reported to for the fluorescence-based detection of organophosphate pesticides. According to the working principle, the nano-system is used in conjugation with Rhodamine B. Under normal condition, the fluorescence of Rhodamine B is quenched by the metallic system. But in presence of organophosphate pesticides, the fluorescence is recovered. This happens because Ag-Au bimetallic system has higher affinity towards organophosphate pesticides than rhodamine B. The turn-on fluorescence response is corelated to the amount of organophosphate present for it quantitative detection [53]. Similarly, metal ions and even gas can be detected with the help of bimetallic biosensors. For instance Hg(II) can be detected by electrochemically by using Au-Pt nanocomposite modified glassy carbon electrode. Bimetallic Au-Pt was modified with organic nanofiber. This inorganic-organic nanocomposite was then utilized for the modification of glassy carbon electrode. The interaction of Hg(II) with the electrode cause variation in electrochemical behaviour, which is used to detect the target analyte voltametrically [54]. On the other hand, ZnO nanowires decorated with Au-Pt bimetallic nano-system was used as a gas sensor for H2S [55]. Similarly, Pd-Au bimetal decorated SnO2 nanosheets based sensor is suitable for formaldehyde and acetone [56].

Therefore, form the above discussion it is clear that a wide range of bimetal-based systems are available which can be used effectively for the detection and estimation of wide range of analytes.

4. Future Aspect

Comprehensive statement regarding Bimetallic nanocrystals has make us luminous that they are exceptionally worthier than those of mono-metallic nanoparticles in connection with their physicochemical as well as optoelectronic properties. It is the proficiency of generation of such materials which decides their crystal parameters (viz lattice strain, defects, crystal structures etc.) as well as kinetic and thermodynamic stability which enhances their electronic coupling/transition properties with itself. Thus, chemical reduction technique is found to be one of the most expedient process which includes both aqueous and non-aqueous methodologies. Since tailoring is the best possible way to determine stoichiometric ratio or epitomizing thickness of core shell structure, therefore implementation of aqueous method is considered to be much greener, as here, water is replacing ecologically unfavourable long chained hydrocarbon reagents. Integration of list of applications of bimetallic nanocrystals may need to accommodate all the application (e.g., drug delivery, bio-imaging, antimicrobial activity, LEDs, sensors etc.), but, since in this chapter we are primarily focusing upon sensors, we are extensively focused upon sensing ability of these metamaterials. Bimetallic nanocrystals have validation not only in terms of chemical sensing, but they can act as bio-sensors as well as bio-analytical sensors too. Now-a-days, deterioration of human health is a common phenomenon and the cause can be anything, like cholesterol, anomalous glucose level, damaged cells or DNA/RNA etc. Therefore, the underlying ambition is to find a kit which has the upbringing of identifying minute fluctuation in such sectors, which is only possible when advanced materials like bimetallic nanocrystals are used in such arrangements. Thence, application of bimetallic nanoparticles in a sensing device is of utmost necessary as they can act as a multitasking sensor material which can be utilized in every domain of human life, starting from regular house hold works to biomedical machineries in modern day warfare. Even the chemical/biological weapons can deprive human health by altering their genetic sequencing, hence immediate detection of malfunctioning is required. Once and for all, bimetallic nanocrystals as sensors, as high end advanced metamaterial, are must needed requirement of this hour due to their capability of get tuned in optoelectronic behaviour as per prerequisite condition and hence can be used as soul material in various sensing devices.

Acknowledgement

Jejiron M. Baruah is grateful to FIRST scheme, North Lakhimpur College (Autonomous) for financial assistance (Ref No.: NLC/RC/MRP/2020/01).

References

[1] Stephanie, R., Kim, M. W., Kim, S. H., Kim, J. K., Park, C. Y., & Park, T. J. (2021). Recent advances of bimetallic nanomaterials and its nanocomposites for biosensing applications. *TrAC Trends in Analytical Chemistry*, 135, 116159.

[2] Wang, A., Liu, X. Y., Mou, C. Y., & Zhang, T. (2013). Understanding the synergistic effects of gold bimetallic catalysts. *Journal of catalysis*, 308, 258-271.

[3] Mandal, R., Baranwal, A., Srivastava, A., & Chandra, P. (2018). Evolving trends in bio/chemical sensor fabrication incorporating bimetallic nanoparticles. *Biosensors and Bioelectronics*, 117, 546-561.

[4] Rajeev, R., Datta, R., Varghese, A., Sudhakar, Y. N., & George, L. (2021). Recent advances in bimetallic based nanostructures: Synthesis and electrochemical sensing applications. *Microchemical Journal*, 163, 105910.

[5] Gilroy, K. D., Ruditskiy, A., Peng, H. C., Qin, D., & Xia, Y. (2016). Bimetallic nanocrystals: syntheses, properties, and applications. *Chemical reviews*, 116(18), 10414-10472.

[6] Ghosh, Chaudhuri, R., & Paria, S. (2012). Core/shell nanoparticles: classes, properties, synthesis mechanisms, characterization, and applications. *Chemical reviews*, 112(4), 2373-2433.

[7] Jana, N. R., Gearheart, L., & Murphy, C. J. (2001). Evidence for seed-mediated nucleation in the chemical reduction of gold salts to gold nanoparticles. *Chemistry of materials*, 13(7), 2313-2322.

[8] Jana, N. R., Gearheart, L., & Murphy, C. J. (2001). Seeding growth for size control of 5− 40 nm diameter gold nanoparticles. *Langmuir*, 17(22), 6782-6786.

[9] Jana, N. R., Gearheart, L., & Murphy, C. J. (2001). Wet chemical synthesis of silver nanorods and nanowires of controllable aspect ratioElectronic supplementary information (ESI) available: UV–VIS spectra of silver nanorods. See http://www. rsc. org/suppdata/cc/b1/b100521i. *Chemical Communications*, (7), 617-618.

[10] Sun, Y., Yin, Y., Mayers, B. T., Herricks, T., & Xia, Y. (2002). Uniform silver nanowires synthesis by reducing AgNO3 with ethylene glycol in the presence of seeds and poly (vinyl pyrrolidone). *Chemistry of Materials*, 14(11), 4736-4745.

[11] Chi, M., Wang, C., Lei, Y., Wang, G., Li, D., More, K. L., ... & Stamenkovic, V. R. (2015). Surface faceting and elemental diffusion behaviour at atomic scale for alloy nanoparticles during in situ annealing. *Nature communications*, 6(1), 8925.

[12] Ahmadi, M., Behafarid, F., Cui, C., Strasser, P., & Cuenya, B. R. (2013). Long-range segregation phenomena in shape-selected bimetallic nanoparticles: chemical state effects. *Acs Nano*, 7(10), 9195-9204.

[13] Serpell, C. J., Cookson, J., Ozkaya, D., & Beer, P. D. (2011). Core@ shell bimetallic nanoparticle synthesis via anion coordination. *Nature chemistry*, 3(6), 478-483.

[14] Skrabalak, S. E., Chen, J., Sun, Y., Lu, X., Au, L., Cobley, C. M., & Xia, Y. (2008). Gold nanocages: synthesis, properties, and applications. *Accounts of chemical research*, 41(12), 1587-1595.

[15] Skrabalak, S. E., Au, L., Li, X., & Xia, Y. (2007). Facile synthesis of Ag nanocubes and Au nanocages. Nature protocols, 2(9), 2182-2190.

[16] Yang, X., Yang, M., Pang, B., Vara, M., & Xia, Y. (2015). Gold nanomaterials at work in biomedicine. *Chemical reviews*, 115(19), 10410-10488.

[17] Yavuz, M. S., Cheng, Y., Chen, J., Cobley, C. M., Zhang, Q., Rycenga, M., ... & Xia, Y. (2009). Gold nanocages covered by smart polymers for controlled release with near-infrared light. *Nature materials*, 8(12), 935-939.

[18] Zeng, J., Zhang, Q., Chen, J., & Xia, Y. (2010). A comparison study of the catalytic properties of Au-based nanocages, nanoboxes, and nanoparticles. *Nano letters*, 10(1), 30-35.

[19] Skrabalak, S. E., Au, L., Lu, X., Li, X., & Xia, Y. (2007). Gold nanocages for cancer detection and treatment. *Nanomedicine*, 2(5), 657-668.

[20] DeSantis, C. J., Weiner, R. G., Radmilovic, A., Bower, M. M., & Skrabalak, S. E. (2013). Seeding bimetallic nanostructures as a new class of plasmonic colloids. *The Journal of Physical Chemistry Letters*, 4(18), 3072-3082.

[21] Weiner, R. G., DeSantis, C. J., Cardoso, M. B., & Skrabalak, S. E. (2014). Diffusion and seed shape: intertwined parameters in the synthesis of branched metal nanostructures. *ACS nano*, 8(8), 8625-8635.

[22] Weiner, R. G., & Skrabalak, S. E. (2015). Metal Dendrimers: Synthesis of Hierarchically Stellated Nanocrystals by Sequential Seed-Directed Overgrowth. *Angewandte Chemie*, 127(4), 1197-1200.

[23] Lim, B., & Xia, Y. (2011). Metal nanocrystals with highly branched morphologies. *Angewandte Chemie International Edition*, 50(1), 76-85.

[24] Wang, Y., Camargo, P. H., Skrabalak, S. E., Gu, H., & Xia, Y. (2008). A facile, water-based synthesis of highly branched nanostructures of silver. *Langmuir*, 24(20), 12042-12046.

[25] Pan, M., Sun, H., Lim, J. W., Bakaul, S. R., Zeng, Y., Xing, S., ... & Chen, H. (2012). Seeded growth of two-dimensional dendritic gold nanostructures. *Chemical Communications*, 48(10), 1440-1442.

[26] HianáTeo, H. (2010). 3D dendritic gold nanostructures: seeded growth of a multi-generation fractal architecture. *Chemical communications*, 46(38), 7112-7114.

[27] Wu, H. L., Chen, C. H., & Huang, M. H. (2009). Seed-mediated synthesis of branched gold nanocrystals derived from the side growth of pentagonal bipyramids and the formation of gold nanostars. *Chemistry of Materials*, 21(1), 110-114.

[28] Wang, C., Xu, C., Zeng, H., & Sun, S. (2009). Recent progress in syntheses and applications of dumbbell like nanoparticles. *Advanced materials*, 21(30), 3045-3052.

[29] Walther, A., & Muller, A. H. (2013). Janus particles: synthesis, self-assembly, physical properties, and applications. *Chemical reviews*, 113(7), 5194-5261.

[30] Personick, M. L., & Mirkin, C. A. (2013). Making sense of the mayhem behind shape control in the synthesis of gold nanoparticles. *Journal of the American Chemical Society*, 135(49), 18238-18247.

[31] Xia, Y., Xia, X., & Peng, H. C. (2015). Shape-controlled synthesis of colloidal metal nanocrystals: thermodynamic versus kinetic products. *Journal of the American Chemical Society*, 137(25), 7947-7966.

[32] Marks, L. D., & Peng, L. (2016). Nanoparticle shape, thermodynamics and kinetics. *Journal of Physics: Condensed Matter*, 28(5), 053001.

[33] Wang, Y., He, J., Liu, C., Chong, W. H., & Chen, H. (2015). Thermodynamics versus kinetics in nanosynthesis. *Angewandte Chemie International Edition*, 54(7), 2022-2051.

[34] Wang, D., & Li, Y. (2011). Bimetallic nanocrystals: liquid-phase synthesis and catalytic applications. *Advanced Materials*, 23(9), 1044-1060.

[35] Liu, X., Wang, D., & Li, Y. (2012). Synthesis and catalytic properties of bimetallic nanomaterials with various architectures. *Nano Today*, 7(5), 448-466.

[36] Pozun, Z. D., Rodenbusch, S. E., Keller, E., Tran, K., Tang, W., Stevenson, K. J., & Henkelman, G. (2013). A systematic investigation of p-nitrophenol reduction by bimetallic dendrimer encapsulated nanoparticles. *The Journal of Physical Chemistry C*, 117(15), 7598-7604.

[37] Ferrando, R., Jellinek, J., & Johnston, R. L. (2008). Nanoalloys: from theory to applications of alloy clusters and nanoparticles. *Chemical reviews*, 108(3), 845-910.

[38] Peng, Z., & Yang, H. (2009). Designer platinum nanoparticles: Control of shape, composition in alloy, nanostructure and electrocatalytic property. *Nano today*, 4(2), 143-164.

[39] Li, Q., Wu, L., Wu, G., Su, D., Lv, H., Zhang, S., ... & Sun, S. (2015). New approach to fully ordered fct-FePt nanoparticles for much enhanced electrocatalysis in acid. *Nano letters*, 5(4), 2468-2473.

[40] Zou, L., Fan, J., Zhou, Y., Wang, C., Li, J., Zou, Z., & Yang, H. (2015). Conversion of PtNi alloy from disordered to ordered for enhanced activity and durability in methanol-tolerant oxygen reduction reactions. *Nano Research*, 8, 2777-2788.

[41] Chen, H., Li, Y., Zhang, F., Zhang, G., & Fan, X. (2011). Graphene supported Au-Pd bimetallic nanoparticles with core-shell structures and superior peroxidase-like activities. *Journal of Materials Chemistry*, 21(44), 17658-17661.

[42] Zhang, X., Wei, M., Lv, B., Liu, Y., Liu, X., & Wei, W. (2016). Sensitive colorimetric detection of glucose and cholesterol by using Au@ Ag core–shell nanoparticles. *RSC advances*, 6(41), 35001-35007.

[43] Tee, S. Y., Teng, C. P., & Ye, E. (2017). Metal nanostructures for non-enzymatic glucose sensing. *Materials Science and Engineering*: C, 70, 1018-1030.

[44] Cao, S., Zhang, L., Chai, Y., & Yuan, R. (2013). Electrochemistry of cholesterol biosensor based on a novel Pt–Pd bimetallic nanoparticle decorated graphene catalyst. *Talanta*, 109, 167-172.

[45] Yao, J., Xie, Z., Zeng, X., Wang, L., & Yue, T. (2022). Bimetallic Eu/Fe-MOFs ratiometric fluorescent nanoenzyme for selective cholesterol detection in biological serum: Synthesis, characterization, mechanism and DFT calculations. *Sensors and Actuators B: Chemical*, 354, 130760.

[46] El-Sheikh, S. M., Osman, D. I., Ali, O. I., Shousha, W. G., Shoeib, M. A., Shawky, S. M., & Sheta, S. M. (2021). A novel Ag/Zn bimetallic MOF as a superior sensitive biosensing platform for HCV-RNA electrochemical detection. *Applied Surface Science*, 562, 150202.

[47] Hu, M., Zhu, L., Li, Z., Guo, C., Wang, M., Wang, C., & Du, M. (2021). CoNi bimetallic metal–organic framework as an efficient biosensing platform for miRNA 126 detection. *Applied Surface Science*, 542, 148586.

[48] Wasfi, A., Awwad, F., Qamhieh, N., Iratni, R., & Ayesh, A. I. (2021). Real-time nucleic acid detection via field-effect transistor sensors based on graphite oxide decorated with trimetallic nanocluster of gold, silver, and platinum. *New Journal of Physics*, 23(10), 103041.

[49] Hosseini, M. (2020). Fast and selective whole cell detection of Staphylococcus aureus bacteria in food samples by paper based colorimetric nanobiosensor using peroxidase-like catalytic activity of DNA-Au/Pt bimetallic nanoclusters. *Microchemical Journal*, 159, 105475.

[50] Prakash, O., Sil, S., Verma, T., & Umapathy, S. (2019). Direct detection of bacteria using positively charged Ag/Au bimetallic nanoparticles: A label-free surface-enhanced Raman scattering study coupled with multivariate analysis. *The Journal of Physical Chemistry C*, 124(1), 861-869.

[51] Lu, C., Tang, L., Gao, F., Li, Y., Liu, J., & Zheng, J. (2021). DNA-encoded bimetallic Au-Pt dumbbell nanozyme for high-performance detection and eradication of Escherichia coli O157: H7. *Biosensors and Bioelectronics*, 187, 113327.

[52] Lu, X., Tao, L., Song, D., Li, Y., & Gao, F. (2018). Bimetallic Pd@ Au nanorods based ultrasensitive acetylcholinesterase biosensor for determination of organophosphate pesticides. *Sensors and Actuators B: Chemical*, 255, 2575-2581.

[53] Luo, Q., Lai, J., Qiu, P., & Wang, X. (2018). An ultrasensitive fluorescent sensor for organophosphorus pesticides detection based on RB-Ag/Au bimetallic nanoparticles. *Sensors and Actuators B: Chemical*, 263, 517-523.

[54] Gong, J., Zhou, T., Song, D., Zhang, L., & Hu, X. (2010). Stripping voltammetric detection of mercury (II) based on a bimetallic Au− Pt inorganic− organic hybrid nanocomposite modified glassy carbon electrode. *Analytical chemistry*, 82(2), 567-573.

[55] Liu, Y., Zhu, L. Y., Feng, P., Dang, C., Li, M., Lu, H. L., & Gao, L. (2022). Bimetallic AuPt alloy nanoparticles decorated on ZnO nanowires towards efficient and selective H2S gas sensing. *Sensors and Actuators B: Chemical,* 367, 132024.

[56] Li, G., Cheng, Z., Xiang, Q., Yan, L., Wang, X., & Xu, J. (2019). Bimetal PdAu decorated SnO2 nanosheets based gas sensor with temperature-dependent dual selectivity for detecting formaldehyde and acetone. *Sensors and Actuators B: Chemical*, 283, 590-601.

Chapter 5

Green Synthesis of Silver-Zinc Oxide and Silver-Titanium Dioxide Bimetallic Nanoparticles

Rutuja Palaskar
Darshana Dhokane
Emmanuel Sebastian
Sonu Mehata
Aditi Chavan
and Balaprasad Ankamwar*
Bio-Inspired Materials Research Laboratory, Department of Chemistry,
Savitribai Phule Pune University (Formerly University of Pune),
Ganeshkhind, Pune, India

Abstract

Nanotechnology is a rapidly developing field of science that deals with the synthesis and further investigation of various nanomaterials. Over the past ten years, bimetallic nanoparticle research has expanded dramatically, mostly due to the use of heterogeneous catalysts in their synthesis. Recently, there has been a surge in interest in using nanoparticles for bimetallic materials. Even in the case of an aggregate of two metal characteristics, bimetallic nanoparticles would exhibit distinct characteristics because of the synergistic effect of metals. Owing to the unique features of ZnO, TiO_2 and Ag nanoparticles, Ag/ZnO and Ag-TiO_2 Nanoparticles is the most remarkable bimetallic nanoparticles.

* Corresponding Author's Email: sb180305@gmail.com.

In: Bimetals
Editors: Pranjal Saikia, Jayanta Kumar Sarmah and Parikshit GogoiName
ISBN: 979-8-89113-496-6

Studies have shown that the production of biogenic nanoparticles is more desirable than traditional methods, which are more expensive and environmentally harmful. Because of their antibacterial and antifungal properties, Ag/ZnO and Ag-TiO_2 nanoparticles are widely used in a variety of industries.

Keywords: bimetallic nanoparticle, Silver-Zinc oxide, Silver-Titanium dioxide bimetallic particles, antibacterial, photodegradation

1. Introduction

1.1. Nanoparticles

The synthesis and evolution of different nanomaterials are at the heart of the newly-emerging scientific field of nanotechnology. The size range of nanoparticles is 1–100 nm, and they have better qualities than bulk material. Because of their high surface to volume ratio, nanoparticles are more affordable, more energy-efficient, and smaller in both size and shape. They are also lighter, faster, and able to pass through tiny pores [1]. The knowledge of physics, chemistry, biology, materials science, health sciences, and engineering is applied in the field of nanotechnology. It has numerous uses in practically every branch of human and scientific study [2]. It has been demonstrated that the conventional physio-chemical method, which synthesizes nanoparticles by combining the physical and chemical methods, is more expensive, requires specialized equipment, and uses dangerous chemicals that pose a variety of biological risks [3]. In order to mitigate these biological risks and steer clear of hazardous chemical usage, scientists have created more advanced techniques that can be applied in two different ways. Using microscopic organisms like yeast, fungus, and bacteria is the first method [4]. It was discovered that the synthesis of nanoparticles using microorganisms was more laborious and necessitated significant steps for maintaining cell culture and intracellular synthesis with additional purification steps. The second method, referred to as "Green synthesis" or "Biogenic synthesis," involves the use of plants. Green synthesis, also known as biosynthesis, is more advanced than physio-chemical methods because it is less expensive, less harmful to the environment, and less toxic [5]. Proteins found in plants serve as capping agents [6]. Proteins are biomolecules with minimal toxicity and degradability [7]. This is the reason that research has found biogenic synthesis of nanoparticles to be more appealing than conventional methods, which are more costly and not environmentally friendly [8]. Green synthesis nanoparticles demonstrated a variety of industrial uses, including medicines, therapies, and pharmaceuticals [9]. Following years of work on green synthesis, nanobiotechnology becomes clear as researchers study how to create nanomaterials from biomolecules found in living things, particularly plants, such as proteins, lipids, secondary metabolites, and metals [10]. Metals have been carefully identified among biomolecules for the synthesis of environmentally friendly nanomaterials. Because of their heterogeneous catalyst synthesis, bimetallic materials have recently attracted increased attention in the application of nanoparticles [11].

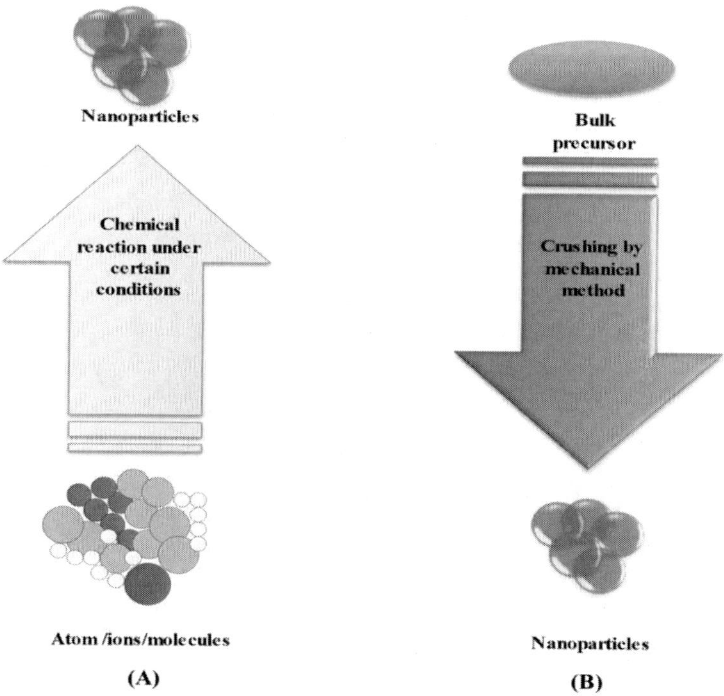

Figure 1. Synthesis method of nanoparticle (A)Bottom-up, (B) Top-down approach [13].

Two main Approaches for the synthesis of nanoparticles:

1. Top-Down Approach
2. Bottom-Up Approach

Using a top-down strategy, the bulk materials trimmed to get the nanoscale particle, while in the bottom-up particle is produced on an atomic scale which subsequently makes up a large nanostructure employing various reactions (Figure 1) [12].

1.2. Bimetallic Nanoparticles

Two distinct metal atoms are combined to form a single nanoparticle known as a bimetallic nanoparticle, or BNP. Because of the synergistic effect of metals, BNPs would display unique features even in the case of an aggregation of two metal characteristics. Monometallic nanoparticles (MNPs) are preferred over bimetallic nanoparticles (BNPs) because of their superior biologic, optical, and catalytic properties (Figure 3). The characteristics of bimetallic nanoparticles are determined by the arrangement of metal atoms. The redox potential of metal ions and the use of reducing agents significantly impact the configuration of BNPs, including heterostructure, alloyed, intermetallic, subclusters, multishell, cluster in cluster, and multishell [14, 15]. Because of its large surface area, nanocomposite exhibits results with greater functionality and unique properties (Figure 2) [16].

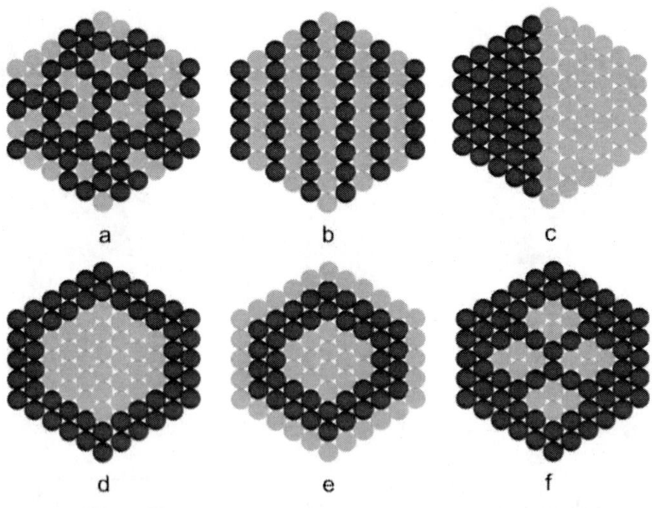

Figure 2. Types of bimetallic nanoparticles: (a) alloyed, (b) intermetallic, (c) subclusters, (d) core-shell, (e) multi-shell core-shell, and (f) multiple core materials coated by a single shell material. Yellow and purple spheres represent two different kinds of metal atoms [17].

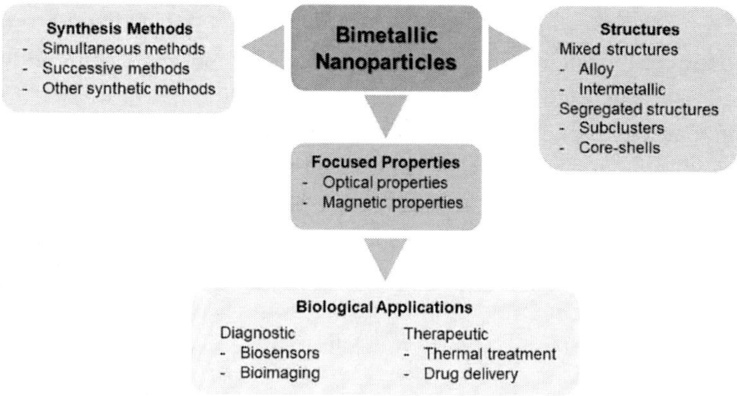

Figure 3. The Synthesis and properties of Bimetallic Nanoparticle for Biological Application [17].

1.2.1. Silver-Zinc oxide Nanoparticles (Ag-ZnO N)

Zinc oxide (ZnO), one of the metal oxides, is a semiconductor metal that is biocompatible and has recently attracted attention in the field of nanomaterial research. Zinc oxide nanoparticles (ZnO NPs) are attractive because of their low cost, ease of synthesis, stability, safety for the environment and human health, as well as their capacity for selective targeting, photocatalysis, and antibacterial activities. Another metal oxide that is widely used in nanomaterials research is silver (Ag). Silver nanoparticles, also known as Ag NPs, are stable, non-toxic, and have antibacterial and photosensitizing effects. Bimetallic nanoparticles are a type of nanocomposite that has drawn attention from researchers. Because of the characteristics of both ZnO and Ag nanoparticles, Ag-ZnO Nanoparticles (Ag-ZnO Nps) are one of the exceptional bimetallic materials. Ag-ZnO nanoparticles exhibit enhanced properties related to photocatalytic performance and antibacterial activity. Ag-ZnO nanoparticles' enhanced properties and activity make them highly applicable in the industrial, medical, and environmental domains [14, 11, 18]. Scientists are interested in the bimetallic nanoparticles Ag-ZnO nanoparticles because of their many potential uses. Ag-ZnO nanoparticles that have been engineered can carry out two different functional tasks. They have a high value capacity as a photocatalyst [20], photodynamic therapy [19], and therapeutic agent for treating a variety of illnesses. Ag-ZnO nanoparticles have a great deal of potential and can meet the demanding requirements of semiconductor applications [21].

1.2.2. Ag-TiO2 Nanoparticles (Ag-TiO_2 NPs)

In contrast to additional bimetallic nanocomposites, the additional characteristics of Ag-TiO_2 nanocomposites, in addition to those of TiO_2 and Ag nanoparticles alone, make them intriguing. Ag is well-known for its use in biological and chemical sensing, while TiO_2 has attracted more attention in the fields of photocatalysis and solar cell applications. The nanocomposite of Ag-TiO_2 increases the Photophysical catalytic properties [22]. TiO_2 NPs are playing an increasingly significant role in the pharmaceutical and cosmetics industries due to their ability to protect skin from UV radiation. Additionally, TiO_2 NPs exhibit notable antibacterial activity [23].

2. Green/Biogenic Synthesis of Silver-Zinc Oxide Nanoparticles (Ag-ZnO) and Silver-Titanium Dioxide (Ag/TiO_2) Nanoparticles

2.1. Silver-Zinc Oxide Nanoparticles (Ag-ZnO NPs)

Ag-ZnO nanoparticles were formed by utilizing *Tetradenia riperia* leaf extract employing a green synthesis technique [24]. The effects of temperature, precursor concentrations, and pH levels on the synthesis of Ag-ZnO nanocomposite (Ag-ZnO NCs) have been investigated in this literature. The effects of pH, precursor concentrations, and temperature were examined in order to improve Ag-ZnO nanocomposite synthesis. Temperature had an impact on the green synthetic nanoparticles' nucleation and growth. Particle growth is favored by low temperatures, while nucleation of small particles is favoured by high temperatures [25, 26]. This study examined three different temperatures: 30°C, 60°C, and 80°C [24]. The final result showed that when the reaction temperature rises, the particle size decreases. This establishes that a temperature increase is the reason for the high metal ion nucleation in comparison to nuclei developing [27, 28]. At high temperatures, the molecule's increased kinetic energy causes precursors to be consumed more quickly in order to create nuclei [26]. The kinetic energy of nanoparticles decreases at low temperatures, leading to the formation of crystals that exhibit Oswald ripening characteristics [29]. Elevated temperatures lead to the formation of more uniformly sized particles and an increase in the surface activity of nuclei, which in turn promotes nuclei to aggregate and collide [27]. Larger particle sizes are also formed at temperatures above 80°C, according to research [25, 27]. The size of the Ag-ZnO nanocomposite is affected by the precursor salt concentration. This was investigated using Ag-ZnO nanocomposite that was doped with silver ions at concentrations of 1 mM, 3 mM, 5 mM, and 8 mM. Ag-ZnO nanocomposite is composed of two phases: nuclei growth and nuclei creation. When the concentration of silver ions rises from 1 to 8 mM, the Ag-ZnO nanocomposite's size decreases. This could be because smaller-sized nuclei are formed at high concentrations of precursor salts due to a rapid rate of ion nucleation [27]. These findings indicate that more nuclei formed lead to a larger concentration of precursor salts, which in turn causes the development of the particle and the agglomeration of the nuclei [30].

The effects of pH optimization on the synthesis of green nanomaterials were studied. The charge of phytochemicals, which function as reducing and capping agents for nanoparticles, is affected by pH variations [27]. The formation and shape of nanoparticles are affected by changes in pH. Tetradenia riperia leaf extract, which was examined in slightly basic and acidic conditions, was used to create the Ag-ZnO nanocomposite. The pH of the extract from T. riperia leaves dropped from 5.10 to 4.46 with the addition of a metal precursor. This can be due to the phytochemicals in *Tetradenia riperia* releasing H+ ions, which are then oxidized insight of metal precursor ions [31]. The phenomenon exhibiting low pH may be caused by the chemical reaction between metal ions and phenol, which reduces metal ions and oxidizes phenol to release hydrogen ions. Because of its low pH, or acidity, it inactivates the phytochemical that caps and reduces the extract's reducing agent and prevents the Ag-ZnO nanocomposite from synthesizing [27, 30]. The conclusion drawn from this literature is that the production of nanoparticles is observed at high pH levels (7 to 8). Analogous findings were reported in other literature [27], where copper nanoparticles from Azadirachta indica extract were generated at pH 6.6, while inhibition in the production of nanoparticles was noted at pH 4.7. High pH was also effective in promoting agglomeration and the production of large-sized nanoparticles during the nanoparticle formation process [26]. Therefore, pH ranges of neutral to slightly alkaline may be optimal for the bio-method's production of small-sized nanoparticles. Through a thorough review of the literature, the simple and environmentally safe synthesis of Ag-ZnO nanocomposites from a variety of plant sources and their parts was described.

Table 1. Green synthesis, characterization and application of Ag-ZnO nanoparticles

Sr No.	Bimetallic	Source	Procedure	Shape and Size	Application	Reference
1	Ag-ZnO NPs	*Tetradenia riperia (TR)* leaves	0.1 M Zinc nitrate hexahydrate +1 mM Silver nitrate +0.1 M Sodium hydroxide + TR leaves extract. Synthesizing Ag-ZnO nanoparticles with different concentration of silver nitrate 1%, 3%, 5%, and 8%.	Spherical and 8% Ag was 12.3 nm	Antimicrobial activity for water disinfection	[24]
2	Ag-ZnO NPs	*Calendulla officinalis*	0.1 M Zinc nitrate hexahydrate+0.01 M Silver nitrate added dropwise to *C. officinalis* extract.	Spherical and 5-72 nm	Antibacterial activity used in areas such as medicine, agriculture, and pharmaceutical industries.	[32]
3	Ag-ZnO NPs	*Azadiracta indica*	Silver nitrate (0.008 mol/L) + Zinc nitrate (0.05	Nanoflowers,1.3-2.3 µm	Antibacterial & antifungal activities	[11]
		Neem extract	mol/L) +2M Sodium hydroxide+ *Neem extract*	Nanoflower made of petal which made up of nanorods (width size 35-80 nm)	exhibited by cotton fabric which can be used in health sector and medical textile in research laboratory.	
4	Ag-ZnO NPs	*Withana somnifera* root	0.5 mmol of Silver nitrate+ ZnO nanopowder suspension 0.5mm ZnO NPs+ DD. H2O+*W.somnifera* root extract	Spherical and ~20nm	Synthesized Ag-ZnO NCs for the reduction of textile dyes act as a nanocatalyst to remediate various environmental issues	[33]

Sr No.	Bimetallic	Source	Procedure	Shape and Size	Application	Reference
5	Ag-ZnO NPs	*Macrotylom a uniflorum* leaf extract	*M. uniflorum* leaf extract +3M Zinc sulfate +0.05M of Silver Nitrate+ Sodium hydroxide	Spherical and 91.17nm	Ag-ZnO showed antimicrobial effect which is potent agent for biological application	[34]
6	Ag-ZnO NPs	Potato Peel Extract	Fine powder of potato peels+ Zinc nitrate hexahydrate + 1 wt% silver nitrate	Spherical Ag on ZnO flower and 15-25nm	Ag-ZnO nanocomposite is great photocatalyst for the degradation of MB.	35]
7	Ag-ZnO NPs	*Excoecaria agallocha* leaf extract	Ultrasound- assisted green synthesis	Spherical and triangular shape and Ag~20nm, ZnO~9 0-100nm	Antimicrobial activity which can be used in antioxidant and anti-inflammatory activity	[36]
8	Ag-ZnO NPs	*Trigonella foenum graecum* leaf	Hydrothermal method	Ag-spherical nanoparticles and ZnO- rods and plates,75nm	Antibacterial, Antifungal, Antioxidant and photocatalytic property	[37]
9	Ag-ZnO NPs	*Valeriana officinalis L.* root extract	Valeriana officinalis L. root extract+0.03 M aqueous solution of the $AgNO_3$+ZnO NPs	Spherical and 40–50 nm	Reusable catalyst for the reduction of organic dyes in a very short time	[38]
10	Ag-ZnO NPs	*Ocimumen uiflorum*(Tulsi)	1 mM zinc nitrate hexahydrate+ (0.5%, 1.0% and 2.0%) silver nitrate added to Tulsi extract	Spherical and 0.5%, 25 nm to 35 nm 1.0%, 15 nm to 22 nm 2.0%, 50 nm to 60 nm	The antimicrobial and dye degradation activity	[39]
11	Ag-ZnO NPs	*Thymus vulgaris*	0.5M Zinc nitrate +Silver nitrate +10M Sodium hydroxide solution	Spherical and ~5nm	Photocatalytic activity	[40]

Table 1. (Continued)

Sr No.	Bimetallic	Source	Procedure	Shape and Size	Application	Reference
12	Ag-ZnO NPs	Turmeric Root	Crude turmeric root powder+ zinc nitrate+ silver nitrate (1, 3, 5, 7 mol %)	Spherical and 20-30nm	Biodiesel synthesis	[41]
13	Ag-ZnO NPs	*Moringa oleifera* leaf extract	*M.oleifera* leaf extract + 0.01M Silver nitrate +0.1M Zinc sulphate heptahydrate	Spherical and 46-66nm	Agriculture (Impact on wheat plant growth)	[16]
14	Ag-ZnO NPs	*Prunus cerasifera leaf extract*	Zinc acetate dihydrate + Silver nitrate + P.cerasifera leaf extract	Size of 30ml is 60–101.98, 50ml is 89.44–101.98, 70ml is 63.25– 120, and 90ml is 60–126.49 nm	Pollutant degradation and invitro biocidal activity	[42]
15	Ag-ZnO NPs	*Oak fruit hall (Jaft)*	Jaft extract +0.M Silver nitrate and 0.1M Zinc acetate	Spherical (crave like structure) and less than 21nm	Photocatalyst, dye degradation	[43]

From overall literature, plant sources used for Ag-ZnO nanocomposite mentioned are *Tetradenia riperia* [24], *Calendulla officinalis* [32], *Azadiracta indica* [11], *Withania somnifera root* [33], *Macrotyloma uniflorum* leaf extract [34], Potato Peel extract [35], *Excoecaria agallocha* leaf extract [36], *Trigonella foenum graecum* leaf [37], *Valeriana officinalis L.* root extract [38], *Ocimum tenuiflorum* (Tulsi) [39], *Thymus vulgaris*, [40], Turmeric root [41], *Moringa oleifera* leaf extract [16] *Prunus cerasifera* leaf extract [42], *Oak fruit hall (Jaft)* [43], *Murraya koenigii* and *Zingiber offcinale* [44]. Some of the other biological sources used for Silver-Zinc Oxide synthesis are mentioned in Table 1.

2.2. Silver -Titanium Oxide Nanoparticle (Ag-TiO_2 NPs)

Studies of green nanoparticle and bimetallic NPs synthesis mediated by plant extract have attracted a lot of attention lately because they are economical, environmentally friendly, and use a small amount of dangerous chemicals in the synthesis of inorganic materials (Figure 4).

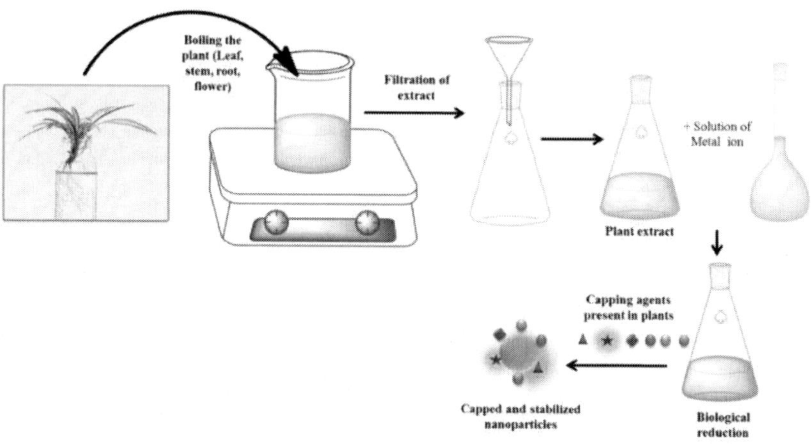

Figure 4. Schematic representation of green synthesis of nanoparticles [13].

Ag-TiO_2 nanocomposites were synthesized by M. Jayapriya et al. using the peel of the Beta vulgaris plant. They started by preparing the peel extract of Beta vulgaris, and then they carried out the synthesis. Essentially, they used extract to create TiO_2 NPs first, and then they added $AgNO_3$, the precursor for Ag NPs, and the same extract once synthesis calcination was completed [45].

During the environmental synthesis of bimetallic Ag-TiO_2 nanocomposite (NCs), an aqueous extract of A. haussknechtii flowers was utilized as a stabilizing/capping agent and a reducing agent. In order to create Ag NPs, they first added $AgNO_3$ to the flower extract, then added TiO $(OH)_2$ solution, and in less than five minutes, they added extract to the first solution once more. Lastly, they centrifuged the mixture [46]. Ag-TiO_2's production, identification, and assessment of its photocatalytic activity were documented by Muhammad Saeed et al. TiO_2 and silver nitrate were employed as precursors, and *Azadirachta indica* leaf extract was utilized as the reducing agent to produce TiO_2 catalysts with loadings of 2, 4, 6, and 8% of silver [After the Ag-TiO_2 NCs were made using sonicated *Origanum majorana* leaf extract, their biological activities were examined. act. They started by making the leaf extract, and then they used titanium (IV) isopropoxide (TTIP) and $AgNO_3$ solution as precursors to create the composite. As a result of surface plasmon resonance (SPR), the remedy's color changed to grey, signifying the production of Ag-TiO_2 nanocomposite, which was reported. The synthesized composite was annealed at 400°C.

2.3. The Plausible Mechanism for Formation of Ag–TiO_2 NCs (Figure 5)

The following equations describe how TTIP hydrolyzes to Ti $(OH)_4$ in aqueous conditions. Ti $(OH)_4$ is often not stable, therefore it would undergo condensation to generate amorphous hydrous oxide precipitates (TiO_2xH2O).

$$Ti(OH)_4 + 4H_2O \text{ (hydrolysis} \dashrightarrow\text{) } Ti(OH)_4 + 4ROH \quad (1)$$

$$Ti(OH)_4 \text{ (condensation} \dashrightarrow\text{) } TiO_2xH_2O + (2 - x)H_2O. \quad (2)$$

The antioxidant properties and condensation processes of *Origanum majorana* leaf extract were caused by the presence of different hydroxyl groups.

The following equation demonstrates how $AgNO_3$ swiftly split into Ag+ and NO_3 ions in the aqueous solution at the same time:

$$AgNO_3 \text{ Hydrolysis} \dashrightarrow Ag+ + NO_3^-$$

Reducing phytochemical consequently bind to and cap the Ag+ ion to create the stable nanoparticles. Carnosol, ursolic acid, carsonic acid, lithospermic acid, and biomolecules found in the leaf extract of *Origanum majorana* are the principal organic components responsible for the synthesis of Ag-TiO_2 NCs [48].

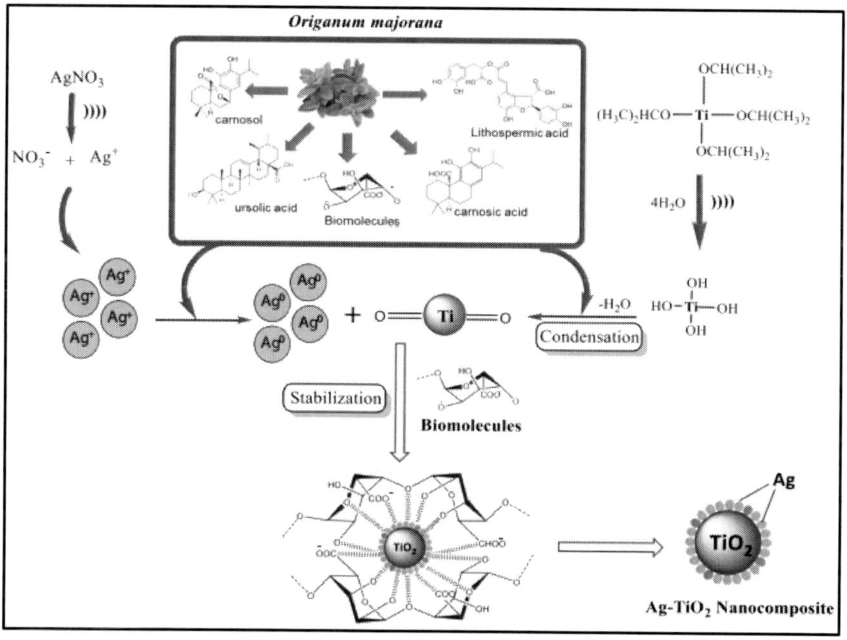

Figure 5. Plausible mechanism for the formation of Ag–TiO_2 NCs using *Origanum majorana* leaf extract [48].

Using the sol-gel method and chemical reduction, gamier leaf (*Uncaria gambir Roxb*) extract was utilized to create the Ag-TiO_2 composite. In order to create control Ag-TiO_2 nanocomposite TiO_2 NPs, they first made the titanate sol by calcining the titanate solution at 500°C for two hours. Ag NPs were created by combining 4% gambir leaf extract with triethanolamine in a molar ratio of 2:2:1, or 0.1M $AgNO_3$:0.5M [49]. Using jasmine flower extract as a basis, titanium dioxide nanoparticles were created during the green synthesis of AC/Ag-TiO_2 nanocomposites. After that, prepared TiO_2 NPs were mixed with AC/Ag composite to create AC/Ag@ TiO_2 nanocomposites, which were then baked for three hours at 300 ºC [50]. The plant extract *Mirabilis Jalapa* served as a reducing agent during the Ag-TiO_2 nanocomposite's formation. Initially, the Ag nanoparticles were made on their own. The $AgNO_3$ and extract solution were then combined with the titanium isopropoxide solution. Two hours of calcination were spent at 450°C. The green synthesis of Ag/TiO_2 using Mirabilis jalapa plant extract has been studied; this is a straightforward, inexpensive, non-toxic, environmentally friendly, and efficient method of using Mirabilis jalapa plant extract. The plant extract functioned as a reducing agent and stabilizer during the production of nanoparticles [51]. Ag-TiO_2 nanocomposite was produced in an environmentally friendly manner with the aid of *Euphorbia heterophylla* leaf extract, and the material's strong catalytic activity for the reduction of various dyes in water was also investigated. Ag NPs were first produced by extracting leaves, and the development of Ag NPs was indicated by a change in color from whitish to dark brown. They found that the conversion of Ag+ into Ag(0) NPs takes twenty minutes. A standard Ag-TiO_2 nanocomposite synthesis involved adding 20 mL of the plant extract dropwise to 40 mL of a well-mixed $AgNO_3$ (0.003 M) and 1.0 g of Degussa P-25 TiO_2 solution while stirring continuously at 60°C for two hours. After 20 minutes, the solution's color changed from yellow to black due to surface plasmon resonance excitation, indicating the formation of the nanocomposite. The bimetallic combination needed to be reduced for roughly 45 minutes [52]. The aqueous extract of Acacia nilotica was used as a reductant in the green synthesis of Ag combined TiO_2 nanoparticles (Ag/TiO_2), and its potential as an antibacterial plus against cancerous cells was evaluated. In order to produce the Ag-TiO_2 nanocomposite, they first prepared the leaf extract from Acacia nilotica. Additionally, the created nanocomposite is annealed for three hours at 400°C [53]. For the synthesis of Ag, fresh lemon fruit (C. Limon) was crushed, filtered, and used as an extract. Ag-TiO_2 and Au/TiO_2 hybrid composites were made by colloidally depositing as-prepared Ag nanoparticles on commercial P25 TiO_2 [54]. Turmeric extract made by adding 6.8 grams of powdered turmeric to 100 milliliters of D/W, boiling, and filtering the mixture. centrifuged for 10 minutes at 250 rpm and 9000 rpm. Sol gel was used to prepare the Ag-TiO_2 nanocomposite. [55] In order to prepare TiO_2 and Ag/TiO_2, P.C. Nethravathi employed a green combustion approach. Silver nitrate served as a dopant, titanium tetrachloride as an oxidizer, and muskmelon juice as a fuel for combustion. The sample was calcined at 500° C to remove the impurities after the fabricated composite was placed into a furnace that had been preheated to 400°C. This caused the vigors to further burn [56]. Extract from leaves of *Eucalyptus globulus L.* made with various combinations of solvents. Pure silver nanoparticles were created using $AgNO_3$ as a precursor and Eucalyptus globulus extract as a reducing agent in order to compare green and conventional methods [57]. Padina tetrastromatica (seaweed) extract, titanium isopropoxide, and $AgNO_3$ solution were used by P Jagadeeswaran et al. to create a biogenic composite. The resultant solution was first heated to 110°C, then to 200°C, and finally to 500°C for annealing [58]. Pranus cerasus is used in the synthesis of Ag-TiO_2. $AgNO_3$ is added to the TiO_2 solution first, followed by the dropwise addition of Pranus cerasus and a 12-hour stirring period. The final product was annealed at 450°C under Ar gas after being cleaned with water and ethanol [59].

The characteristics of nanomaterials were altered by using a range of plant extracts containing particular amounts and combinations of biomolecules. Because of this, the various plant sources used to create nanomaterials exhibit varying morphological and structural changes in addition to varying antibacterial efficacies. This encourages a more thorough investigation of plant species and the synthesis of nanomaterials from a variety of plant species and plant constituents. Alkaloids, flavonoids, proteins, tannins, terpenoids, saponins, and polyphenols are just a few examples of the phytochemicals, or active ingredients, that function as capping and reducing agents in the plant extract. Reducing agents encourage reactivity with metal ions (a naturally weak base), while capping agents work with nuclei to stop nanoparticle aggregation. Because of their potential, phytocompounds can be used to create Ag-TiO_2 Bimetallic nanoparticles in an environmentally friendly way.

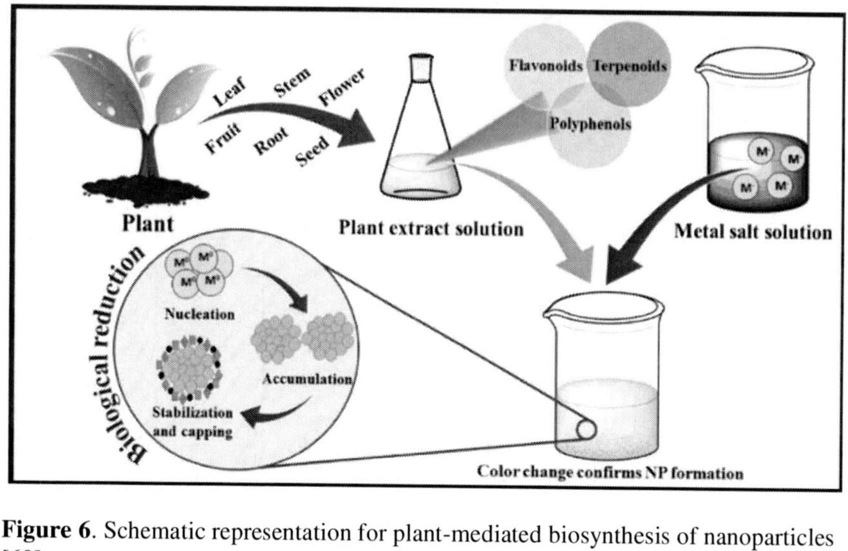

Figure 6. Schematic representation for plant-mediated biosynthesis of nanoparticles [60].

Table 2. Green synthesis, characterization and application of Ag/TiO_2 nanocomposites

Sr.No.	Bimetallic Nanoparticle	Source	Synthesis parameters	Shape and Size	Applications	References
1)	Ag/TiO_2	*Azadirachta indica* leaves extract	Leaves extract+ AgNO + TiO_2	FCC(Ag) Spherical & uniform (TiO_2)	Photocatalytic activity	[47]
2)	Ag/TiO_2	*Euphorbia heterophylla* leaf extract	Leaf extract +$AgNO_3$ soln.+Degussa P-25 TiO_2	FCC(Ag) 24 nm for both NPs	Catalytic activity in dyes reduction	[52]
3)	Ag/TiO_2	*A. haussknechtii* flower extract	Flower extract+ $AgNO_3$ soln then added $TiO(OH)_2$ soln	Spherical and 36.99+_12.03 nm	Antiplanktonic, antibiofilm, antiswarming motility & antiquorum sensing	[46]
4)	Ag/TiO_2	*C. limon* fruit extract	Colloidal soln of Ag+ TiO_2	Quasi-Spherical & 2-3 nm	Photocatalytic activity	[54]
5)	Ag/TiO_2	*Beta vulgaris* peels extract	$TiO(OH)_2$+Peels extract then added $AgNO_3$ soln+Peels extract	Spherical shaped & 40nm	Antibacterial activity Catalytic degradation of dyes & organic pollutants	[45]
6)	Ag/TiO_2	*Mirabilis jalapa* plant extract	*Mirabilis jalapa* extract+$AgNO_3$ +TTIP	Spherical shaped & 160 nm	-	[51]
7)	Ag/TiO_2	*Origanum majorana* leaf extract	*Origanum majorana* leaf extract+TTIP then added $AgNO_3$ soln.+ again leaf extract	Spherical shaped Ag NPs on TiO_2 & 25-50 nm	Antimicrobial activity Antioxidant activity	[48]
8)	Ag/TiO_2	*Acacia nilotica* leaves extract	Tetrabutyl orthotitanate +ethanol+HNO_3+*Acacia nilotica* leaves extract+$AgNO_3$	Spherical & 20-40 nm Crystallite size of Ag/TiO_2:9 nm	Antimicrobial activity Anticancer activity	[53]

Sr.No.	Bimetallic Nanoparticle	Source	Synthesis parameters	Shape and Size	Applications	References
9)	Ag/TiO_2	jasmine flower extract	Jasmine flower extract +AC/Ag powder+TiO_2 soln	flake, sphere and one dimensional (1D) needles & 24 to 36 nm	Antibacterial activity Photocatalytic Activity	[50]
10)	Ag/TiO_2	*Uncaria gambir Roxb*	Titanate sol soln+ag sol soln+	-	Photocatalytic Activity	[49]
11)	Ag/TiO_2	Cucumis melo juice	Titanium tetrachloride +muskmelon juice + Silver nitrate	Spherical shape & 9.12 nm	Antioxidant activity Photocatalytic Activity	[56]
12)	Ag/TiO_2	*Eucalyptus globulus* L. Extract	TiO_2 NPs +$AgNO_3$ +*Eucalyptus globulus* L. Extract	non-defined, non-uniform surface, almost round particle + 14.53 nm to 14.3 nm	Antimicrobial activity	[69]
13)	Ag/TiO_2	*Padina tetrastromatica* (seaweed)	titanium (IV) isopropoxide + $AgNO_3$+ *Padina tetrastromatica* (seaweed) extract	25 nm.	-	[58]
14)	Ag/TiO_2	*Pranus cerasus*	Silver nitrate +TiO_2 solution+ *Pranus cerasus*	Ag NPs (10-20) nm & TiO_2 (30-40) nm	Photodegradtion of organic pollutant in water	[59]

A variety of plant extracts with distinct biomolecule combinations and specific concentrations were used to modify the properties of nanomaterials. As a result, different plant sources that are used to create nanomaterials show different changes in structure and morphology in addition to different levels of antibacterial efficacy. This motivates scientists to look into new plant species more thoroughly by promoting the synthesis of nanomaterials from a variety of plant sources and plant components, such as leaves, roots, stems, and seeds. A few phytochemicals or active ingredients that are present in plant extract and act as reducing and capping agents are alkaloids, flavonoids, proteins, tannins, terpenoids, saponins, and polyphenols. Reducing agents improve reactivity by working with metal ions, which are inherently weak bases. Capping agents work with nuclei to prevent nanoparticle aggregation. Using a variety of plant species and plant components, the potential of phytocompounds enables their use in this environmentally friendly process of turning silver and zinc salts into Ag-ZnO nanoparticles [24] (Figure 6).

The brief format of Silver -Titanium dioxide nanoparticles synthesis, applications, source, size, shape has been mentioned in Table 2.

3. Characterization of Bimetallic Silver-Zinc Oxide Nanoparticles (Ag-ZnO) and Silver-Titanium Oxide (Ag/TiO_2) Nanoparticles

3.1. UV-Visible Spectroscopy

The first method to verify synthesized nanoparticles was UV-visible absorption spectroscopy. The unique Surface Plasmon Resonance (SPR) characteristic of noble metals is tracked by UV-visible spectroscopy and significantly influences the amount of incident light that is absorbed. Ag-ZnO NPs were synthesized using a green method using seed extract from *Ocimum tenuiflorum* (Tulsi), with three different concentrations of Ag: 0.5%, 1.0%, and 2.0%. This process was recorded by double beam for 0.5%, 1.0%, and 2.0% of Ag-ZnO NPs, respectively, in the UV-visible spectrophotometer analysis range of 350 to 700 nm. The absorbance of prepared Ag-ZnO NPs increased as the concentration of silver nitrate was increased from 0.5% to 1.0%, with an intense peak observed at 415 and 427 nm. However, if the concentration of silver nitrate is further increased to 2.0%, the absorbance decreases and the band gap increases. Ag-ZnO NPs at varying ionic concentrations were optimized for 0.5%, 1.0%, and 2.0%. The approximate values of Ag-ZnO NPs with optical band gaps of 0.5%, 1.0%, and 2.0% are 2.99 eV, 2.90 eV, and 3.12 Ev [39]. Ag and Ag-ZnO SPR characteristics were used to synthesize Ag-ZnO nanoparticles (NPs) using floral extract from Calendulla officinalis. UV-visible spectroscopy of the extract reveals absorption due to band gap modification in the visible region between 400 and 500 nm. Additionally, the shift in absorption spectra caused by Ag deposition on ZnO's surface provides some insight into the Ag-ZnO interface, and the XRD patterns support these findings [32].

Additionally, the shift in absorption spectra caused by Ag deposition on ZnO's surface provides some insight into the Ag-ZnO interface, and the XRD patterns support these findings [32]. In order to produce Ag-ZnO nanocomposites in a green manner, *Tetradenia riperia* leaf extract was employed. The UV-Vis absorption spectra of the 1%, 3%, 5%, and 8% doped ZnO nanocomposites showed absorption peaks at 397, 380, 377, and 368 nm, which were attributed to surface plasmon resonance [24]. Green synthesis from *Azadirachta indica* Ag-ZnO nanoflowers (AZO) show a notable red shift of the absorption peak at 365 nm, whereas Ag-ZnO nanoparticles show a UV absorption peak at 355 nm. Peak shifting could be brought on by the increase in ZnO particle size that happens when it is mixed with nano Ag. The visible absorbance for AZO at 450 nm [11] confirms the presence of Ag in Ag-ZnO nanoflowers. UV-visible spectrum of wavelength 300-520 nm shows the characterization peak for ZnO at 330 nm and for Ag-ZnO bimetallic NPs it was recorded in the region of 330 nm and 366-379 nm. Numerous characterization peaks in the 360–410 nm range attest to the production of Ag–ZnO bimetallic NPs [16]. In the UV-Vis absorption band, synthesized Ag-ZnO NCPs showed both strong and modest peaks at 395 and 550 nm. The strong interfacial coupling between ZnO NPs and Ag NPs is responsible for these peaks [21]. Ag-ZnO nanocomposites show a peak at 372 nm from facile synthesis using Macrotyloma Uniflorum leaves, and a strong Surface Plasmon Resonance peak at 365 nm from ZnO. Ag is present on the surface of Zn, as indicated by a new peak at 415 nm [34].

Initially, Ag-TiO_2 nanocomposites made with extract from *Mirabilis jalapa* were examined using UV-visible spectroscopy. A wavelength shift was also noted, indicating oxidation following the addition of plant extract, and a sharp absorption band at 460 nm was observed, indicating the dispersion of particles. The surface plasmon resonance (SPR) band of silver nanoparticles showed absorption band characteristics, visible between 390 and 500 nm. Round-shaped nanoparticles were confirmed at 300–700 nm, suggesting a single surface plasmon resonance. Broadening the absorbance is said to have an impact on the size and form of the NPs [51]. Ag-TiO_2 nanoparticles were made using extract from leaves of E. heterophylla. Surface plasmon resonance was found to influence the absorption spectra of an Ag-TiO_2 nanocomposite when metallic nanoparticles were present, as detected by UV-Vis spectroscopy. In contrast to the Ag NPs' spectra, the reaction's color changed from pale to dark, indicating the formation of nanostructures and the appearance of surface plasmon resonance (SPR) between 250 and 350 nm, as confirmed by UV-visible spectroscopy. The produced nanostructures were evaluated using UV-visible, and it was found that even after a month, there was little variation in the shape, location, or symmetry of the absorption peak, indicating product stability [52]. Following the observation of a color shift in the solution, which suggested the continuous formation of Ag nanoparticles, UV-Vis spectra of the manufactured nanoparticles were obtained [49]. The UV-Vis spectra of the synthesized Ag-TiO_2 nanocomposite were examined by Diksha Bharadwaj et al. [48]; the results showed that TiO_2 exhibited an absorption band around 380 nm, while Ag NPs on the surface of NCs exhibited an absorption band around 570 nm. The UV-Vis spectra of monometallic NPs, floral extract, and Ag-TiO_2 NCs were examined by Mehran Alavi et al. They found that those varied in terms of absorbance. Ag NPs showed peaks at 382 and 397 nm, while TiO_2 NPs showed peaks at 253 nm. Ag-TiO_2 NCs showed an absorbance shift to 496 nm. On the other hand, the floral extract did not show any clear peaks [46]. The broad absorption peak of the AC/Ag-TiO_2 nanocomposite was observed at 608 nm. The combination of silver and titanium ions shows how closely the orbits of Ag and TiO_2 are related. The strong absorption spectra were observed around the Fermi level due to the strong hybridization between the O 2p and Ti 3d bands and the silver orbitals. The optical bandgap of the sample under investigation was calculated using the Tauc plot. The AC, AC/Ag, and AC/Ag-TiO2 nanocomposite exhibit optical band gap values of 2.75 eV, 2.25 eV, and 1.99 eV, in that order [50].

3.2. X-Ray Crystallography

X-ray powder diffraction (XRD) is a quick analytical method that can be used to determine the phase of a crystalline material and provide details on unit cell dimensions. XRD can be used to investigate the phase variations of monometallic NPs as their diameter increases. Since the diffraction pattern of the physical mixtures differs from that of the BNPs and has merging lines from the two different monometallic NPs, X-ray diffractograms (XRD) can be used to demonstrate the existence of bimetallic particles rather than a mixture of monometallic particles [61].

The hexagonal wurtzite segment for ZnO NPs in the biogenic mediated synthesis of Ag-ZnO NCs by seed extract of *Ocimum tenuiflorum* (Tulsi) revealed XRD peaks at 31.87°, 34.56°, 36.47°, 47.7°, 57.32°, 63.20° 67.20°, 68.34°, 69.68°, 73.10°, and 78.34° for pure ZnO-NPs, corresponding to planes to 100, 002, 101, 102, 110, 103, 200, 112, 201, 004, and the other 3 peaks for Ag at ZnO at 38.4, 44.5, and 67.5 correspond to the established silver revealing cubic segment for Ag corresponding to (JCPDS file no 2-109) [39]. Ag-ZnO NPs were synthesized using Calendulla officinalis floral extract. The Ag-ZnO NPs nano-dimensions with a 17.98 nm particle size were shown by their XRD pattern, which included 2θ values at 31.92°, 34.53°, 36.46°, 44.65°, 56.62°, 64.87°, and 68.08°. Crystallite size is calculated using the Scherrer method. However, because dopants prevent the lattice strain from being counted, precise size cannot be determined using this method [32]. By using With reference to JCPDS card No. 04-0783, XRD peak studies using *Tetradenia riperia* leaf extract in the green synthesis of Ag-ZnO nanocomposites reveal the face-centric cubic (FCC) shape of the Ag nanoparticles, indicating the presence of Ag NP in the nanocomposite. Ag-ZnO composites' decreased crystallinity and particle size are caused by articles' peak intensity rising with concentration and ZnO nanoparticles' peak intensity falling [24]. Ag-ZnO NPs from Azadirachta indica (neem extract) were found to have cubic structure in the Ag and wurtzite hexagonal structure in the ZnO, according to XRD peaks that matched the corresponding JCPDS card numbers [11]. Ag-ZnO nanoparticles XRD pattern revealed crystals with a hexagonal wurtzite shape found consistent with the Joint Committee on Powder Diffraction Standards card no. and an average crystallite size of 10.02 nm calculated by Scherrer's equation (D = ×.[K λ β/ cosθ ×Å) [42]. Ag-ZnO nanoparticles' XRD pattern revealed crystals with a hexagonal wurtzite shape, which was found to be consistent with card number of the Joint Committee on Powder Diffraction Standards. Scherrer's equation (D = ×.[K λ β/ cosθ ×Å)] yielded an average crystallite size of 10.02 nm [42]. Ag-ZnO nanocomposite growth is referred to as an XRD pattern based on the face- centered-cubic (FCC) structures distinctive diffractions correspond to respective JCPDS Card no. 04-0783 .38.1°, 44.3°, 64.6°, and 77.4° are different diffraction peaks by XRD spectrum indexed to the (111), (200), (220), and (311) planes, with favoured orientation in the (111) direction. This XRD diffraction peaks confirmed Ag and ZnO nanocomposite synthesis [21].

The spectrum was reported by Raghavendra K. et al. and is referenced in JCPDS files 36-1451 and 04-0783. All of the synthetic composites made from Macrotyloma Uniflorum leaves have a face-centered cubic structure (FCC) and are polycrystalline in nature. The ZnO spectra show an intense peak at 37.98o, which is equivalent to (002), and the Ag-doped ZnO spectra show that this peak remains unchanged, indicating that Ag does not cause any defects in the ZnO crystallites [34]. The green synthesis of Ag-ZnO NCPs using *essential ginger oil* shows spectrum reveals 5 diffraction peaks labeled at (31.8°, 34.1°, 36.1°, 56.6°, and 67.7°), were indexed to the (100), (002), (101), (110), and (112) planes.The preferred orientation of ZnO NPs is represented by the (100) direction index. Additionally, based on the characteristic diffractions of the face-centered-cubic (FCC) structured corresponding to (JCPDS Card no. 04-0783), the XRD pattern determines the extension of Ag-ZnO NCPs. Several diffraction peaks with a preferred orientation in the (111) direction can be seen in this XRD spectrum, which are indexed to the (111), (200), (220), and 77.4^, respectively. It is reported that the synthesis NCPs include Ag and ZnO based on these XRD diffraction peaks [21].

Using an XRD, the phase analysis and crystallinity of the synthesized Ag-TiO_2 were assessed. At 25°, 36°, 48°, and 55° in the TiO_2 XRD pattern, there were notable and comparatively well-defined peaks that matched the anatase structural phase diffraction planes (1-0-1), (0-0-4), (2-0-0), and (1-0-5) (JCPDS 21-1272), in that order. The Ag-TiO_2 spectra show two additional peaks at 238° and 44°, which are consistent with the face-centered cubic unit cell of Ag [47]. The crystalline sizes of TiO_2 and Ag-TiO_2 were found to be approximately 9 nm and 17 nm, respectively, based on XRD analyses of silver doped titanium dioxide (Ag-TiO_2) nanoparticles mediated by Acacia nilotica. The XRD pattern verifies the anatase TiO_2 (JCPDS 21-1272) tetragonal structure. The absence of a silver peak suggests that the metal is evenly distributed as a tiny cluster throughout the TiO2 matrix [53]. The crystalline structure and particle size of Ag-Cu, Cu-Ag, Ag-TiO_2, and TiO_2 NCs were determined by XRD analysis. XRD analysis was used to estimate the crystalline structure and particle size. Additionally, they found additional unmarked peaks close to typical peaks, which may have been caused by stabilizers already in place capping the surface of NPs. Ag-TiO_2 and TiO_2-Ag NCs were measured for diffraction intensity at 10-80°. In these samples, four prominent peaks were identified at 24 (TiO_2-Ag), - 30 (Ag-TiO_2), 38, 49, and 78. These peaks correspond to the planes of Ag (111), TiO_2 (001), TiO_2 (200), and Ag (311), in different orders. Based on the anatase form of TiO_2 NPs and the cubic crystal structure of Ag, these peaks can be indexed [46]. The additional diffraction peaks at 2 theta values 32.27 and 46.94, indexed to (1 1 1) and (2 0 0) planes, match to face-cantered cubic geometry of silver nanoparticles (JCPDS no. 2-109), which verifies the successful doping of Ag on the surface of TiO_2 nanoparticles, according to crystallographic studies of the nanocomposite produced by green combustion technique. The anatase TiO_2 NPs' tetragonal structure was confirmed by comparing the data to JCPDS Card No. 21-1272. The average crystal size (D) of the produced TiO_2 and Ag-TiO_2 photocatalysts was found to be 9.093 nm and 22.8 nm, respectively using Scherrer's equation [56].

3.3. Fourier-Transform Infrared Spectroscopy

This technique is widely used to determine the vibrational spectra of microscopic molecules attached to nanoparticle surfaces. These molecules are usually generated from substances that function as reducing or stabilizing agents in the reaction solution. They are useful for qualitatively and quantitatively determining the molecular composition of organic molecules in the NPs structure, particularly in hollow-core/porous-shell materials T. Sathish et al. state that Ag-ZnO nanocomposite's capping and reducing functional biomolecules are identified by FTIR. ucing Ag-ZnO nanocomposite. Withania somnifera's main polyphenols, amides, and amino acids can be identified using FTIR in the 1200–1500 cm^{-1} vibrational intensity range [33]. Biogenic mediated synthesis of Ag-ZnO nanocomposite (with varied concentrations of Ag 0.5%, 1.0%, and 2.0%) revealed FTIR spectrum identifies the alcohols, carboxyl, aromatic, and alkyl halide which are considered with reduction mechanism using a seed extract of *Ocimum tenuiflorum* (Tulsi) [39]. The peak at 2974 in the Ag-FTIR spectrum, which corresponds to the C-CH_3 stretching, and the peak at 1346, which corresponds to the CH_2 stretching, indicate that Ag-ZnO NPs were synthesized using the floral extract of Calendulla officinalis. The presence of primary and secondary amines in the synthesized nanoparticles is determined by the adsorption band at 1409 cm^{-1}, which corresponds to the CH_2 vibration [31]. The adsorption band at 1409, which corresponds to the CH_2 vibration, indicates whether primary and secondary amines are present in the synthesized nanoparticles [31]. Plant extracts from Murraya koenigii and Zingiber officinale were used to synthesize Ag-ZnO nanocomposites (NCs). The band that was detected at approximately 1060 cm^{-1} may indicate the van der Waals forces of interaction between Ag and C–C bonds in Ag-ZnO NCs. The presence of functional groups in the easily synthesized Ag-ZnO NCs is determined by their FTIR spectra, which supports this [44]. The potato residue utilized in the Ag–ZnO nanocomposites synthesis exhibits an FTIR band at 1410–1622 cm^{-1}, which is explained by the hydroxide (–OH) groups absorbed at the surface of the samples bending vibrations. The vibrations of the organic residuals attribute weak bands observed at 2364, 1095 and 1041 cm^{-1}. The strongest and most prominent absorption band, which emerged at 530 cm^{-1}, may be associated with the M–O bonds (M=Zn and Ag) [35]. The attachment of functional groups of Excoecaria agallocha leaf extract over the surface of ZnO suggests shifting of peaks towards the lower wavenumber. Biopolymer relevant functional groups, such as polyols, carboxylic, hydroxyl, and terpenoids, may act as reducing agent and stabilizing agent in the synthesis of nanoparticles. On the other hand, peaks below 500 cm^{-1} are associated with the formation of ZnO nanoparticles [36]. The extract of Trigonella foenum graecum leaf, also known as fenugreek leaves, was used in the biological synthesis of Ag-ZnO nanocomposites. The FTIR spectra of this leaf extract show a sharp band at approximately 520 cm^{-1}, which indicates the vibration of the Zn - O bond in the ZnO composition, and weak bands in the range of 900 to 1500 cm^{-1}, which indicate the extract composition on the surface of the Ag-ZnO nanocomposites [37].

M. Jayapriya et al. stated that the B. vulgaris peel extract contains a hydroxyl group (OH), which has shown a stretching frequency of 3440 cm^{-1}. the 400–900 cm^{-1} region of the spectrum of absorption that corresponds to Ti–O–Ti vibration. Ag nanoparticles, OH functional groups, carbonyl group (C--O), stretching vibrations of the C—C aromatic ring, and the C-OH stretching vibrations, respectively, are represented by the vibration bands seen at 1069, 1491, and 3721 cm^{-1}. It was determined from the results that phytoconstituents in B. vulgaris extract have been in charge of generating Ag-TiO_2 nanocomposites [45]. Functional groups from polyphenolics revealed phenolic compounds in plant extract by their presence. Phenolic component can be in charge of the reduction of metal ions & creation of nanocomposite was reported by Frida Octavia Purnomo et al. from FT-IR analysis [51]. The peaks at 3884.64 and 3361.93 cml are typical of the alcohol group's -OH stretching vibrations, whereas the peak at 1627.92 cm^{-1} is typical of the alkenes group's C=C stretching vibrations. According to the oxide structure, 476.42 and 468.70 cm bands [58].

3.4. Morphological Studies

Microscopic techniques like atomic-force microscopy (AFM), scanning electron microscopy (SEM), field emission scanning electron microscopy (FESEM), transmission electron microscopy (TEM) and high-resolution TEM (HRTEM), and analysis have all been used to study the size, morphology, and distribution of nanostructures. SEM and AFM images are inappropriate for understanding the structure of core-shell NPs because they only describe the surfaces. It is too challenging to distinguish the core in SEM images, even though TEM images are especially helpful for understanding the same because they can measure the thickness and distance between the core and shells. Energy dispersive spectrometry (EDS) (TEM and SEM) is an add-on for electron microscopes. Its powerful method for determining the distribution of elements in the materials under study has made the composition of the shell and core visible [62]. The ability of scanning electron microscopes to produce three-dimensional (3D) images that provide information about typology, topography, and makeup is extremely significant in the field of nanotechnology research. It offers fine-grained images of the surface of a sample. This has made it possible for the SEM to magnify images up to 200 times. It is an essential method for particle size measurement and nanomaterial characterization [61].

Biogeneic mediated synthesis of Ag-ZnO NPs (with distinct concentrations of Ag 0.5%, 1.0%, and 2.0%) revealed hexagonal and spherical form of both Ag in ZnO nanocomposites for different concentrations, as demonstrated by TEM analysis using a seed extract of Ocimum tenuiflorum (Tulsi). Various sizes were found based on the concentration: 1.0% showed 15 to 22 nm, 2.0% showed 50 to 60 nm, and 0.5% showed 25 to 35 nm. As 0.5% and 1.0% of Ag-ZnO NPs indicated a reduction in agglomeration but 2.0% of them affected the ZnO morphology and the Ag-ZnO NPs once again shown agglomeration, SEM was used to assess the change that was caused in the morphology of generated samples [39]. Using the floral extract of *Calendulla officinalis*, Ag-ZnO nanoparticles were created. The production of these particles is visible in the method's TEM images [32]. Green synthesis of Ag-ZnO nanoparticles utilising *Tetradenia riperia* leaf extract analysis TEM micrograph of 8% Ag reveals further information on the coexistence of Ag on ZnO where an average particle size of 14.8 nm was determined [24]. Nanoflowers, 1.3 to 2.3 m were produced by the green synthesis of Ag-ZnO flowers (AZO) from *Azadiracta indica*. Nanorods with a width ranging from 35 to 80 nm are used to make nanoflower petals [3]. The spherical form of bimetallic Nanoparticles (NPs) derived from Moringa oleifera through green synthesis is demonstrated by a SEM structural study. Between 46 and 66 nm is the range of nanoparticle sizes [16]. Triangular cages of ZnO NPs evenly distributed and nanowire-shaped Ag-ZnO NCPs are visible in the FESEM image [21]. The reduction in particle size and aggregation observed in the distribution of nanocomposites by SEM could be explained by the interaction of silver nanoparticles with ZnO. Using the linear intercept approach, the average size was found to be roughly 128 nm for ZnO and 90 nm for Ag-ZnO. The findings demonstrated that the nanoparticles are deposited above the rod and have a spherical shape. A green synthetic Ag-ZnO nanocomposite made from the aqueous extract of Jaft has been shown in a FESEM image to have silver and ZnO nanoparticles with the same diameter, or less than 21 nm, and a spherical shape with cave-like features [43]. The morphology of Ag-ZnO nanoparticles made from turmeric root extract was studied using SEM images, which revealed agglomeration and spherical structures of varying sizes. TEM images revealed that the Ag-ZnO nanoparticles were between 20 and 30 nm in size [41]. Silver particles are not incorporated into the ZnO lattice but rather are deposited on the ZnO surface. This is demonstrated by the SEM image of the Ag-ZnO nanocomposite synthesised from an extract of Valeriana officinalis L. root green, which shows diameters of less than 24 and 40–50 nm for Ag and ZnO NPs, respectively. TEM images revealed spherical particles with a size of 36.99 nm, while Mehran Alavi et al. used FESEM to demonstrate spherical, worm-like, and aggregated structures [46].

The FESEM image of the synthesized Ag-TiO_2 nanocomposite revealed the formation of smaller, monodispersed spherical nonentities, possibly due to orientation adhesion and grouping between the Ag and TiO_2 nanoparticles. Furthermore, the development of a spherical-shaped nanocomposite was supported by TEM research, which showed an average diameter of roughly 40 nm [45]. TEM images with low resolution showed a spherical particle measuring 9.12 nm. HRTEM provided the D-spacing values of 0.1954 and 0.2380 for Ag and TiO_2, respectively. It is easy to identify silver on titanium dioxide as a black spot because silver has a heavier atomic weight than titanium. It was confirmed by SEAD [56] that the generated particle was polycrystalline. The Ag-TiO_2 catalyst has irregularly shaped particles, which may be the result of small crystals aggregating, and the SEM image demonstrates that the distribution of silver on the surface of titanium dioxide is not uniform. That some silver particles are too tiny to see at the resolution of the used microscope cannot be completely ruled out, though. The SEM image additionally indicates that the catalyst surface topology remains unchanged following silver metal doping [58].

3.5. Energy–Dispersive X-Ray Spectroscopy

In this test, the surface of one kind of conductive material is struck by an electron beam, which then produces an X-ray beam. Upon scanning the material with an electron beam, an elemental map is generated. Crucial details about the chemical composition of the NPs can be found using an EDX high-resolution microscope [63]. The strongest peaks for zinc, silver, and oxygen ions were observed using an EDX spectrum between 0.1 and 1.5 KeV, which was used to determine the elemental composition [16]. According to EDX, *Ocimum tenuiflorum* (Tulsi) plant seed extract was utilized to create Ag-ZnO NCs via a green method at different Ag concentrations (0.5%, 1.0%, and 2.0%). The synthesized Ag-ZnO NCs (1.0%) showed three distinct peaks at about 0.8 kev, which were caused by the presence of oxygen atoms; two additional peaks occurred at 1.7 kev connected to zinc and 3.1 kev related to silver [39]. The presence of Ag, Zn, and O elements in the samples was revealed by the EDX spectra of Ag-ZnO nanocomposites (NCs), which were synthesized using plant extracts of Murraya koenigii and Zingiber officinale. This result is consistent with the XRD results. Since the typical contamination peak is absent, the synthesized nanomaterials' purity is confirmed [44]. The EDX technique was used to characterize the extract of Trigonella foenum graecum leaf, which was used in the biological synthesis of Ag-ZnO nanocomposites. The resulting EDX spectrum shows a purity percentage of 88.94 for the synthesized Ag-ZnO nanocomposites, along with impurities related to the sample preparation steps. The synthesized Ag-ZnO nanocomposite had a weight percentage of 5.60 for Ag and 86.02 for ZnO [37].

Using the EDX technique, Diksha Bhardwaj et al. reported the elemental composition of Ag-TiO_2 NCs and confirmed the presence of titanium, silver, and oxygen. The fabricated Ag-TiO_2 nanocomposite contained 40.22% titanium, 36.99% silver, and 16.60% silver. In the EDX spectrum of Ag-TiO_2 nanocomposites, Ti, Ag, O, Ca, Cl, and Si are present. Calcium and chloride were extracted from seaweed and used in the nanocomposites. It is evident from the spectra that the composites have been suitably described because there are notable peaks for Ti, Ag, and O [58]. According to Farnosh Tavakol's analysis of Ag-TiO_2-G(G-graphene) elemental composition, Ag, Ti, O, and C are the primary elements. Ag & TiO_2 NPs were also found to be well-placed over the graphene surface [59].

4. Application of Bimetallic Silver-Zinc Oxide (Ag-ZnO) Nanoparticle and Silver-Titanium Oxide (Ag/TiO_2) Nanoparticles

Recent years have seen significant development in bimetallic nanoparticles, which have unique physical, chemical, and biological properties. This has opened the door to a wide range of applications. It is applicable in many different forms because of the distinct mechanical, electrical, chemical, biological, and thermal characteristics that arise from the fusion of two distinct metals in a single nano system. The global agricultural industry is constantly impacted by a variety of pests and microbes that hinder plant growth, leading to problems with food security and agricultural manufacturing failures. In order to treat plant diseases brought on by different phytopathogens, farmers usually employ fungicides or bactericides; however, doing so is toxic and unbalances the environment. Consequently, more work needs to go into developing strategies that will successfully replace these man-made chemicals with ones that are less harmful to the environment and living things. Because of these special characteristics, bimetallic nanoparticles are used as essential tools for the confirmation of viruses, fungi, and phyto-pathogens. They are also used to track various environmental issues like salinity and drought conditions, as well as the concentration of heavy metals and minerals in the soil [1].

The antibacterial and dye degradation properties of pure 0.5%, 1.0%, and 2.0% Ag-ZnO NPs that were made using the green route from Ocimum tenuiflorum (Tulsi) plant seed extract were examined. The synthesized 1.0% Ag-ZnO NCs demonstrated excellent photocatalytic and antibacterial activities in comparison to the previously produced Ag-ZnO NCs. Ag-ZnO NCs' antimicrobial and dye-degrading properties are enhanced by the production of ROS (reactive oxygen species). By reducing oxygen by one electron in this case, Ag particles on the ZnO surface, which promote electron transfer, may enhance the formation of reactive oxygen species [39]. The pharmaceutical, agricultural, and medical industries frequently use environmentally safe Ag-ZnO nanostructures derived from Calendula officinalis flower extracts, which have shown antibacterial activity [32]. Ag-

ZnO nanoparticles, which have good antibacterial activity against Escherichia coli and Staphylococcus aureus, can be biogenically synthesized using Tetradenia riperia leaf extract and used for a number of applications, including water disinfection [24]. Research labs and other healthcare facilities may use cotton fabric with antibacterial and antifungal properties, as demonstrated by the biogeneic synthesis of Ag-ZnO flowers (AZO) from Azadirachta indica (neem extract) [11]. Ag-ZnO that has been biosynthesized from Calotropis gigantea has shown to be a dependable and reusable catalyst in degradation studies [64]. When different dyes are reduced in water, an extract from Valeriana officinalis L. root green synthesis of Ag-ZnO nanocomposite exhibits catalytic activity. Without losing its catalytic activity, this used catalyst can be recycled and recovered numerous times. Making this process more environmentally friendly is part of this methodology [38]. Using an extract of Trigonella foenum graecum leaves, also known as fenugreek leaves, biological synthesis of Ag-ZnO nanocomposites was performed. This plant has antibacterial, antioxidant, and photocatalytic qualities. Compared to S. aureus, the synthetic Ag-ZnO nanocomposites had less of an impact on E. coli in terms of antibacterial activity. *S. aureus* is a Gram positive bacteria, while *E. coli* is a Gram negative. Because of this and the differences in their cell wall composition, it can be concluded that Gram-positive bacteria are more vulnerable to antibacterial agents than Gram-negative bacteria. The cell wall or membrane structure of bacteria demonstrates their capacity to withstand the antibacterial effects of Ag-ZnO nanocomposite.

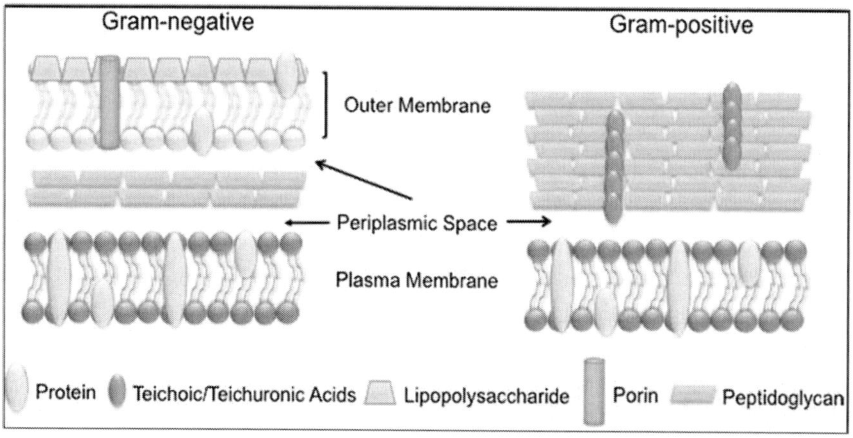

Figure 7. Comparison of bacterial cell wall structure [65].

Gram negative bacteria have been observed to be more susceptible to the effects of ZnO nanoparticles than Gram positive bacteria. The peptidoglycan layer, which makes up the cellular wall of Gram-positive bacteria, is approximately 20 nm thick and is composed of cross-linked straight polysaccharides. That peptidoglycan coating, however, is only 7–8 nm thick on gram negative bacteria. Above it, however, is the lipopolysaccharide layer. Gram negative bacteria have a thicker lipopolysaccharide layer than Gram positive bacteria, which makes them more resistant to nanoparticles [66].

Ag-ZnO nanocomposites have the potential to prevent microbial contamination in food packaging and surgical instrument coatings because of their antibacterial activity. Ag's surface contains functional groups, which are responsible for its antioxidant qualities. Studies have demonstrated that adding Ag to ZnO nanoparticles increases their antioxidant capacity and, because of their synergistic effect, eliminates free radicals. Therefore, in terms of DPPH free radical scavenging, Ag-ZnO nanocomposites exhibit higher antioxidant activity than ZnO nanoparticles. Given that the dye was completely removed in 120 minutes, Ag-ZnO nanocomposites may be highly capable of breaking down malachite green dye [37]. When compared to urea, the application of bimetallic nanoparticles (NPs) derived from Moringa oleifera in agriculture has a markedly positive impact on wheat varieties, as evidenced by improved growth factors (physiological and morphological attributes) [16]. Additionally, bimetallic nanoparticles work incredibly well as antibacterial agents. Bimetallic nanoparticles with antimicrobial properties can combat microorganisms in combination with antibiotics. By rupturing bacterial membranes or releasing reactive oxygen species (ROS), which damage DNA and prevent the bacterium from producing proteins, these nanoparticles can prevent the growth of bacteria [67] (Figure 8).

Ag-TiO_2 nanocomposite exhibits good antibacterial activity at concentrations of 20 g/ml and 25 g/ml, with inhibition zones of 8 mm and 10 mm for S. aureus and P. aeruginosa, respectively [45]. Ag-TiO_2 nanocomposites showed exceptional catalytic activity against methylene blue and rhodamine B dyes; however, as Ag concentration rose, it blocked TiO_2's active center, reducing photocatalytic activity [47]. Ag-doped TiO_2 NPs demonstrated better antiradical activity than Ag & TiO_2 NPs by reducing the viability of DPPH radical scavenging. It proved that the generated photocatalyst could be helpful in solving global environmental issues, such as water pollution, by being used for extensive water treatment [20]. Materials such as silver-modified titanium dioxide have the superior photocatalytic and antibacterial properties that can help address the problem of microorganisms and persistent chemicals contaminating water [21]. It has been discovered that the impact of silver nanoparticles (Ag NPs) is more noticeable for Gram-negative bacteria than for Gram-positive bacteria. This is because Gram-negative bacteria have a thinner cell wall (3–4 nm), whereas Gram-positive bacteria have a cell wall (30 nm), which makes particle penetration more difficult. Ag-TiO_2 materials prepared with the conventional reductive reagent yield high inhibition percentages, which is directly associated with the antibacterial properties of silver. Due to the lack of a UV light source during the experiments, which could have activated the material, only titanium dioxide NPs show low inhibitory activity. However, materials made using plant extract from *Eucalyptus globulus L.* show high percentages of inhibition, up to 100% when microwaves are used to aid in the reaction mechanism [22]. Ag-TiO_2 was used to photocatalytically degrade Acid Orange 7 (AO7).

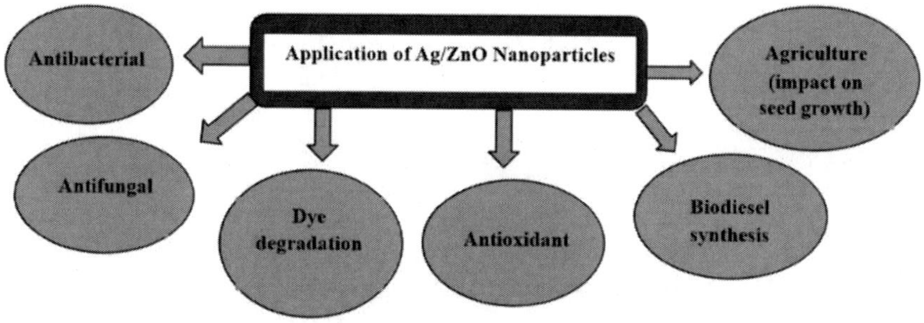

Figure 8. Flowchart showing Application of Ag-ZnO Nanoparticles.

G compound, which showed that there was no AO7 degradation when the dye was the only one exposed to UV light. This proved that the photocatalyst is the only component involved in degradation. to investigate two composites' (Ag-TiO_2 and Ag-TiO_2-G) photocatalytic performance in the presence and absence of graphene. The results show that the remarkable properties of graphene, including its large surface area, electron mobility, and capacity to delay the reintegration of electrons and holes, significantly affect the photodegradation of pollutants [24]. With rate constants of 0.14, 0.18, and 0.055 min^{-1}, respectively, the Ag-TiO_2/PG photocatalyst showed strong photocatalytic activity in the removal of rhodamine B (RhB), methylene blue (MB), and methylene orange (MO) in the photodegradation of organic dyes. To evaluate the Ag-TiO_2/PG photocatalyst's photocatalytic activity at the time of creation, organic dyes and compounds containing aromatic sulfur were photocatalytically oxidized. Under visible light irradiation, a DBT conversion of 83.5% over Ag-TiO_2/PG was attained in 80 minutes, and a BT conversion of 80.1% was reached in the same amount of time. The photocatalytic oxidation of DBT and BT was driven by reactive oxygen species, which are byproducts of photoinduced electron-hole pair reactions with trace amounts of dissolved O_2 and H_2O. By not exposing the photocatalytic oxidation desulfurization process to external oxidants like H_2O_2 or O_2, the explosion risk—which is crucial for industrial applications—was successfully avoided [25] (Figure 9).

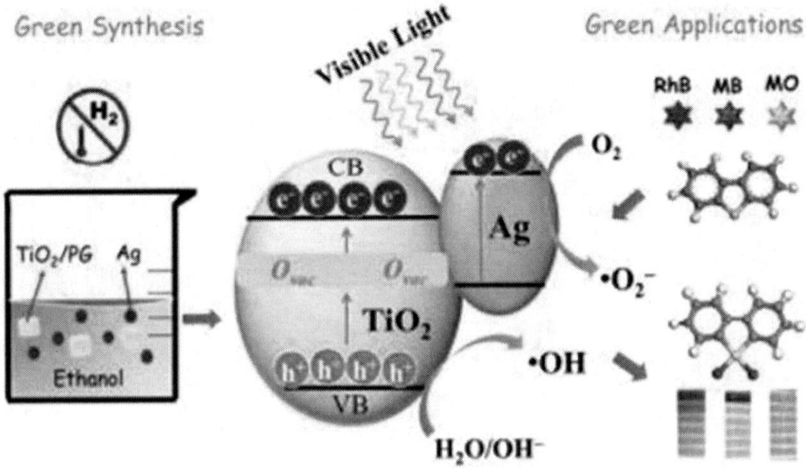

Figure 9. Photocatalytic Degradation of RhB, MB, MO [68].

4.1. Mechanism of AO7 Photodegradation

Series of free radicals trapping experiments were done to understand the photocatalysis of Ag-G compound in Acid Orange 7 (AO7) degradation which indicated that hydroxyl radicals & photoinduced holes are important players in degradation of AO7 [24].

4.2. Mechanism of Antibacterial Activity

Since Ag NPs attack bacteria through multiple pathways, the fundamental mechanism by which they suppress them is still unclear. Despite this, microorganisms do not seem to become resistant to silver. Four different mechanisms allow Ag NPs to exert their antimicrobial effects: (1) adhesion to the cell wall and membrane; (2) penetration into intracellular structures and biomolecules that damage cells; (3) NP-induced cellular toxicity and oxidative stress due to the production of free radicals and reactive oxygen species (ROS); and (4) modification of the signaling pathways involved in cell transduction [21]. There is not just one thing that kills the bacteria (Figure 10). These mechanisms would also be harmful to human cells because lipids, proteins, and DNA are similar biomolecules. However, by using particular ligands and bacterial cell receptors, potential treatments for bacterial infections may be targeted precisely. The multi-target activity caused by NPs makes it ideal to treat and eradicate multi-drug resistant bacteria, even though it is unlikely that they can mount multiple defenses at once [26].

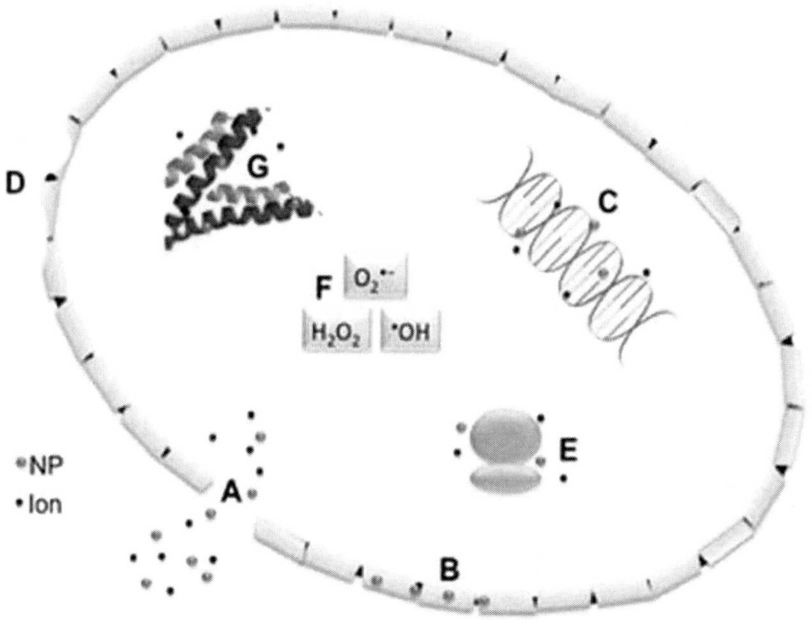

Figure 10. A suggested model illustrating how AgNPs exposed to Gram-negative E. coli cells work. (A) Cell wall disintegration enables intracellular components to exit the cell. (B) AgNPs entering the periplasmic space, starting the membrane's separation from the cytoplasm. (C) AgNPs' interaction with DNA. ROS generation may result from inhibition. Cell pits that develop following exposure (D). (E) Improper DNA function, malformed or suppressed proteins, and inhibition of appropriate ribosome action, all of which result in the creation of ROS. ROS production is (F). (G) How proteins, namely cysteine, interact with one other [65].

AgNPs' bactericidal qualities are largely dependent on the form, size, concentration, and colloidal state of the particles. It appears that stability and biocompatibility are enhanced by smaller AgNP sizes. It does not matter that minute AgNPs interact with cell membranes, alter the lipid bilayer, increase permeability, and eventually cause cell death. UV light and TiO_2 are delivered separately. damaged the outermost layer of LPS in E. Coli cells, but not the peptidoglycans; this resulted in regular creases that resembled groove-like rifts, twisting, and a rough appearance. In addition, cells bulged at both extremities, growing shorter and wider but still resembling rods. These alterations in bacterial morphology indicated some external cellular layer disintegration. Conversely, cells exposed to TiO_2 and UV changed from their typical rod shape to an elliptical, spheroplast-like form. These results suggest damage to the phospholipid layer, LPS, peptidoglycans, and outer membrane. It is interesting to note that bacterial cells had an intact inner membrane and were still alive and well [69].

Conclusion

Ag-ZnO and Ag-TiO_2 nanoparticles are two types of bimetallic nanoparticles that scientists are currently studying and finding interesting. The biological process that produces bimetallic nanoparticles is environmentally safe and serves as a reducing, capping, and stabilizing agent. Future studies can create bimetallic nanoparticles by utilizing the wealth of biological resources present in nature. Several methods are employed to characterize the size, shape, and morphology of bimetallic nanoparticles. By using different sources and carrying out different procedures, different sizes, shapes, and morphologies can be observed, which can be used in a range of applications. Another potential future aspect is using numerous protocols to obtain appropriate size, shape, and morphology. Application of Ag-ZnO nanoparticles in a variety of industrial area, such as waste water cleanup to get rid of microbes and dyes. Ag-ZnO nanoparticles are used in many different fields, including waste water treatment, where their antimicrobial, antifungal, and photocatalytic qualities aid in the removal of dyes and microorganisms. Utilizing Ag-ZnO nanoparticles in agriculture to promote plant growth is one application that can be beneficial to farmers. The biosynthesized Ag-TiO_2 material's bacterial inhibition values were found to be promising for its potential use in systems where bacteria are an issue, such as water potabilization or biologically treated water disinfection. It is possible to substitute phytoconstituents for hazardous compounds without compromising the material's properties by skipping purification steps and removing any remaining harmful agents from the synthesis processes. This enables the use of readily available materials in scenarios requiring close contact with living organisms. Future uses of Ag-ZnO and Ag-TiO_2 nanoparticles made via the biological approach may benefit society without harming the environment.

Author Credentials

Rutuja Palaskar, Darshana Dhokane,- Writing, concept designing, image designing; Emmanuel Sebastian - Literature review; Sonu Mehata, Aditi Chavan- Referencing and Balaprasad Ankamwar*- Validation and Review.

References

[1] Saba, H.; A Review on Nanoparticles: Their Synthesis and Types, Research Journal of Recent Sciences, *Res. J. Recent. Sci.*, 2015, *vol. 4(ISC-2014)*, 1-3.

[2] Godwin, M. A. K.; Mahitha, S.; Balaji, M.; Nanoparticles and Their Applications – A Mini Review, *IJREB,* 2015, *Vol 3 Issue 5*,11- 29.

[3] Geoprincy, G.; Vidhya, S. B. N.; Poonguzhali, U.; Nagendra, G. N.; Renganathan, S.; A Review on green synthesis of Silver Nanoparticles. *Asian. J. Pharm. Clin. Res.*, 2013, *6(1),* 8-12.

[4] Helan, J. C.; Anand, R. L. F. A.; Namasivayam, S. K. R.; Bharani R. S. A., Improved pesticidal activity of fungal metabolite from Nomureaerileyi with chitosan Nanoparticles, *ICANMEET, IEEE:* 2013 387-390.

[5] Vidya, C.; Hiremat, S.; Chandraprabha, M. N.; Antonyraj, L. M. A.; Gopal, I. V.; Jain, A.; Bansal, K., Green synthesis of ZnO Nanoparticles by Calotropis gigantea, *Int. J. Curr. Eng. Technol.*, 2013, *vol 1:* 118-120.

[6] Sangeetha, G.; Rajeshwari, S.; Venkatesh, Green synthesis of Zinc oxide Nanoparticles by Aloe barbadensis miller leaf extract: Structure and optical properties, *R., Mater. Res. Bull.,* 2011,*.46:* 2560– 2566.

[7] Raja, S. K. N.; Anand, R. L. F. A.; Robin, A. T. G.; Helan, J. C.; Arvind, R. S. B., Optimal synthesis of biocompatible bovine serum Nanoparticles-incorporated quercetin (BSA NPS-QT) nano drug conjugate for the controlled release and improved anti oxidativeactivity *Res. J. Pharm., Biol. Chem. Sci.,* 2014, *vol 5*, 478-487.

[8] Raj, L. F. A. A.; Jayalaksmy, E., Biosynthesis and characterisation of Zinc oxide Nanoparticles using root extract of Zingiber officinale, *Orient. J. Chem.*, 2015, *vol 31(1),* 51-56.

[9] Mohanpuria, P.; Rana, N. K.; Yadav, S. K., Biosynthesis of Nanoparticles: technological concepts and future applications, *J. Nanoparticles. Res.,* 2008, *vol 10 (3)*, 507-517.

[10] Paul, A.; Roychoudhury, A., Go green to protect plants: repurposing the antimicrobial activity of biosynthesized silver Nanoparticles to combat phytopathogens, *Nanotechnol Environ Eng*., 2021, *vol. 6*,1–22.

[11] Malobi, S.; Hasmat, K.; Sunirmal, Antimicrobial activity study of Ag-ZnO nanoflowers synthesised from neem extract and application in cotton textiles, J., *Int. J. Nanotechnol*., 2021. *Vol. 18, Nos. 5/6/7/8*, 531-544.

[12] Zhang, D.; Ma, X.; Gu, Y.; Huang, H. and Zhang, G., Green synthesis of metallic nanoparticles and their potential applications to treat cancer, *Front. Chem*. 2020. *8*,799.

[13] Singh, C.; Mehata, A. K.;Priya, V.; Malik, A. K.; Setia, A.; Suseela, M. N. L.; Vikas, G. P.; Samridhi; S. S. K.; Madaswamy, S. M., Bimetallic Au–Ag Nanoparticles: Advanced Nanotechnology for Tackling Antimicrobial Resistance. *Molecules* 2022, *27(20)*,7059.

[14] Maria, E.; Abdul, W.; Abd, U.; Abeer, K.; Amir, A.; Naveed, I. R.; Zia-ur-Rehman, M.; Tahira, S.; Nilofar, M.; Muhammad, I.; Huanyong, L., Plant-Based Bimetallic Silver-Zinc Oxide Nanoparticles: A Comprehensive Perspective of Synthesis, Biomedical Applications, and Future Trends. *Biomed. Res. Int.* 2022, *Volume 2022*, 1-20.

[15] Ferrando, R.; Jellinek, J.; Johnston, R. L., Nanoalloys: From Theory to applications of alloy clusters and Nanoparticles. *Chem. Rev.* 2008, *108*, 845–910.

[16] Maria, E.; Naveed, I. R.; Ziaur-ur-Rehman, M.; Muhammad, I.; Efat, Z.; SyedaSadaf, Z.; Fozia, A.; Mubashir, H.; Muhammad, I.; Nilofar, M.; Asad, A., Responses of bimetallic Ag/ZnO alloy Nanoparticles and urea on morphological and physiological attributes of wheat, *IET Nanobiotechnol.* 2021 *vol.15*, 602– 610.

[17] Pannaree, S.; Yi-Ting, C.; Varadee, V.; Maria, D.; Marquez, T.; Randall, L., Bimetallic Nanoparticles: Enhanced Magnetic and Optical Properties for Emerging Biological Applications *Appl. Sci.* 2018,*8,* 1106.

[18] Efa, K.; Green synthesis and characterization of Ag/Zn bimetallic Nanoparticles using garlic(Allium sativum L) plant extracts for antibacterial applications, [M.S. thesis], ASTU, 2021.

[19] Yavari, H.; Ghasemi, N.; Divband, B.; Rezaei, Y.; Jabbari, G.; Payahoo, S.; The effect of photodynamic therapy and polymer solution containing nano-particles of Ag/ZnO on push-out bond strength of the sealers AH-Plus and MTA Fillapex. *J. Clin. Exp. Dent.*, 2017, *vol. 9(9)*, e1109-e1114.

[20] Ding, C.; Fu, K.; Pan,Y.; Liu, J.; Deng, H.; Shi, J., Comparison of Ag and AgI-modified ZnO as heterogeneous photocatalysts for simulated sunlight driven photodegradation of Metronidazole, *Catalysts*, 2020, *vol.10(9), 1097.*

[21] Asaad, A. M.; Abid, A. M.; Khaldoon, N. A.; Wisam J. A.; Salim, A. A., Photocatalytic Activity of Ag-ZnO Nanocomposites Integrated Essential Ginger Oil Fabricated by Green Synthesis Method, *J Phys Conf Ser* (2021), *1892.*

[22] Indrajit, S.; Masaki, U.; Toyoko, I., Synthesis of -Cyclodextrin-Modified Water-Dispersible Ag-TiO_2 Core–Shell Nanoparticles and Their Photocatalytic Activity *J. Nanosci. Nanotechnol*. 2011, *Vol. 11 (4)* 2011.3764.

[23] Noushin, A.; Keerti, S.; Afroz, M; Bakht, Y. R.; Mohammed, J. A.; Noushad, M., Cost-effective and eco-friendly synthesis of titanium dioxide (TiO_2) nanoparticles using fruit's peel agro-waste extracts: characterization, in vitro antibacterial, antioxidant activities, *Green Chem Lett Rev*, 2019,*12:3,* 244-254.

[24] Stanslaus, G. M.; Revocatus, L. M.; Bart van der, B.; Karoli, N. N. In situ facile green synthesis of Ag–ZnO nanocomposites using Tetradenia riperialeaf extract and its antimicrobial efficacy on water disinfection, *Sci. Rep*. 2022, *12*, Article no. 15359.

[25] Hongyu, L.; Huan, Z.; Jie, W.; Junfu, W., Effect of temperature on the size of biosynthesized silver Nanoparticles: deep insight into microscopic kinetics analysis, *Arab. J. Chem.* 2020, *13(1),* 1011–1019.

[26] Aparajita, V.; Mohan, S. M., Controllable synthesis of silver Nanoparticles using Neem leaves and their antimicrobial activity *J. Radiat. Res. Appl. Sci,*. 2016, *9(1)*, 109– 115.

[27] Niharika, N.; Vijay, D., Green synthesis and characterization of copper Nanoparticles using Azadirachta indica leaves. *Mater. Chem. Phys.* 2018, *213*, 44-45.

[28] Mostafa, M. H.; Khalil, E. H.; Ismail, K. Z.; El-Baghdady, D. M., Green synthesis of silver Nanoparticles using olive leaf extract and its antibacterial activity. *Arab. J. Chem.* 2014, *7(6),*1131–1139.

[29] Satyanarayana, R. A.; Chien-Y. C.; Simon, C. B.; Chien, C. C.; Jiin-Shuh, J.; Cheng-Wei, F.; Hau-Ren, C.; Jung-Chen, W., Synthesis of silverNanoparticles using surfactin: A biosurfactant as stabilizing agent, *Mater. Lett.* 2009, ***63****(15)*,1227–1230.

[30] Vanaja, M.; Paulkumar K.; Baburaja, M.; Rajeshkumar, S.; Gnanajobitha, G.; Malarkodi, C.; Sivakavinesan, M. Annadurai, Degradation of methylene blue using biologically synthesized silver Nanoparticles. *Bioinorg. Chem. Appl.* 2014. *vol.2014,* 8.

[31] Yallappa, S.; Manjanna J.; Sindhe, M. A.; Satyanarayan. N. D.; Pramod, S. N.; Nagaraja, K., Microwave assisted rapid synthesis and biological evaluation of stable copper Nanoparticles using *T. arjuna*bark extract. *Spectrochim. Acta A Mol. Biomol. Spectrosc.,* 2013.*110*, 108–115.

[32] Mamta, S.; Harish, B.; Priyanka, C.; Pankaj, T.; Ankush, C.; Rajesh, K.; Ravichandran, H. K.; Naresh, K. T.; Inhibition of the bacterial growth as a consequence of synergism of Ag and ZnO: *Calendula officinalis* mediated green approach for nanoparticle and impact of altitude, *Inorg Chem Commun.* 2022, *136*, 109131.

[33] Sathish, T.; Chandramohan, D.; Dinesh, K. S.; Rajkumar, S.; Vijayan, V., A facile synthesis of Ag/ZnO nanocomposites prepared via novel green mediated route for catalytic activity, *Appl. Phys.* A 2021, *127*:692.

[34] Raghavendra, K. S.; Malatesh, S. P.; Shivaprasadagouda, P.; Ashok, H. S., Green Synthesis of ZnO and Ag-ZnO Nanoparticles using Macrotyloma Uniflorum: Evaluation of Antibacterial Activity, *Adv. Mater. Lett.* 2021,*12(7)*,21071645.

[35] Fahad A. A.; Abdul, A. A. A.; NabilAl-Zaqri, H.; Alanazi, S.; Amjad, A. A.; Adel, E. M.; Naushad, A., Facile one pot green synthesis of Ag-ZnO Nanocomposites using potato peel and their Ag concentration dependent photocatalytic properties, *Sci. Rep.* 2020 *10*:20229.[D29].

[36] Mujahid, S. K.; Pratik, P. D.; Bhaskar, L. J.; Navinchandra, G. S., Ultrasound-Assisted Green Synthesis of Ag-Decorated ZnO Nanoparticles UsingExcoecaria agallocha Leaf Extract and Evaluation of Their Photocatalytic and Biological Activity, C*hemistry select*, 2020, *5,* 12660–12671.

[37] Zahra, N.; Hamzeh, A.; Saeed, F.; Abdelnaser, M. G., Green synthesis of Ag –ZnO nanocomposites using Trigonella foenum-graecum leaf extract and their antibacterial, antifungal, antioxidant and photocatalytic properties, *Spectro Chim. Acta A Mol. Biomol. Spectrosc,* 2020, *vol.240. 118595.*

[38] Ali, Y. F.; Maryam, B.; Neda, N.; Mahmoud, N., Green synthesis of the Ag/ZnO nanocomposite using Valeriana officinalis L. root extract: application as a reusable catalyst for the reduction of organic dyes in a very short time, *IET Nanobiotechnol.*, 2017, *Vol. 1(6),* 669-676.

[39] Priyanka, P.; Devina, R. P.; Anshu, S.; Pratibha, C.; Poonam, M.; Nehra, S. P., Biogenic mediated Ag/ZnO nanocomposites for photocatalytic and antibacterial activities towards disinfection of water, *J Colloid Interface Sci*, 2020 *563*, 370–380.

[40] Mina, Z.; Keerthiraj, N.; Saad, A.; Yasser, Hussein, E. M.; Abdo, H.; Mohamad, Z.; Qasem, A. D.; Kullaiah, B.; Bananakere, N. C.; Seeram, R.; Xiang Z., Novel Green Biomimetic Approach for Synthesis of ZnO-Ag Nanocomposite;Antimicrobial Activity against Food-borne Pathogen,Biocompatibility and Solar Photocatalysis, *Sci. Rep* (2019) *9*, 8303.

[41] Reddy Y.; Pratibha, S.; Manjunath, K.; Shivanna, M.; Ramakrishnappa T.; Dhananjaya, N.; Nagaraju, Green synthesis of Ag-ZnO Nanoparticles: Structural analysis, hydrogen generation, formylation and biodiesel applications, G. *J SCI-ADV Mater Dev,* 2019, *4*, 4 25e431.

[42] Shaan, B. J.; Khuram, S. A., Foliar-mediated Ag:ZnO nanophotocatalysts: green synthesis, characterization, pollutants degradation, and in vitro biocidal activity, *Green Process. Synth.* 2019, *8*: 172–182.

[43] Mina S.; Ebrahim, S. M.; Ali Ramazani.; Saeid, T. F., Biosynthesis of Ag, ZnO and bimetallic Ag/ZnO alloy Nanoparticles by aqueous extract of oak fruit hull (Jaft) and investigation of photocatalytic activity of ZnO and bimetallic Ag/ZnO for degradation of basic violet 3 dye, *J Mater Sci: Mater Electron*, 2018, *29*, 2806– 2814.

[44] Dominic.; Savio, A. S.; Murugankeerthi.; Sundararajan, M.; Sugumar, S.; Basavegowda L.; Vijaykumar, V.; Aziz, K. R.; *In vitro* cytotoxicity efficacy of phytosynthesized Ag/ZnO nanocomposites using Murraya koenigii and Zingiber offcinale extracts, *Mater. Chem. Phys.* 2021, *272*, 124903.

[45] Jayapriya, M., M. Arulmozhi, Beta vulgaris peel extract mediated synthesis of Ag/TiO2 nanocomposite: Characterization, evaluation of antibacterial and catalytic degradation of textile dyes-an electron relay effect, *Inorg. Chem. Commun*. 2021 *128*, 108529.

[46] Mehran, A.; Naser, K., Antiplanktonic, antibiofilm, antiswarming motility and antiquorum sensing activities of green synthesized Ag–TiO_2, TiO_2–Ag, Ag–Cu and

Cu–Ag nanocomposites against multi-drug-resistant bacteria, *Artif Cells Nanomed. Biotechnol*, 2018 *46:sup3*, 399-413,

[47] Muhammad S.; Majid, M.; Muhammad K.; Khan, K.; Nadia, A.; Sheeba, K.; Muhammad, A.; Asif, N.; Sonia, S.; Azadirachta indica leaves extract assisted green synthesis of Ag-TiO2 for degradation of Methylene blue and Rhodamine B dyes in aqueous medium, *Green Process Synth*,2019,*8*,659-666.

[48] Diksha, B.; Ruby, S.; Green biomimetic synthesis of Ag–TiO_2 nanocomposite using Origanum majorana leaf extract under sonication and their biological activities, *Bioresour. Bioprocess*. 2021,*8(1)*.

[49] Sry Wahyuni.; Syukri.; Syukri, A., Green synthesis of Ag/TiO_2 Nanocomposite Assisted by Gambier Leaf (UncariagambirRoxb) Extract, *J. kim. sains apl,* 2019, *22 (6),* 250-255.

[50] Amalanathan, M.; Parvathiraja, C.; Asma A. A.; Saikh M. W.; Mohammad, A. I.; Enhanced photocatalytic and biological observations of green synthesized AC, AC/Ag and AC/Ag/TiO_2 nanocomposites, 2021.

[51] Frida, O. P.; Sari, S. N.; Sadwika, N. K., Synthesis and Characterization of Ag/TiO_2 Nanoparticles using Mirabilis jalapa Plant Extract, *J. kim. Valensi*, 2021, *Vol 7(1),* 22-27.

[52] Monireh, A.; Mahmoud, N.; Mohammad, S. S., Euphorbia heterophylla leaf extract mediated green synthesis of Ag/TiO_2 nanocomposite and investigation of its excellent catalytic activity for reduction of variety of dyes in water, *J Colloid Interface Sci,* 2016, *462,* 272–279.

[53] Tentu, N. R.; Riyazuddin, P. B.; Naushad, A.; Rais, A. K.; Iftekhar, H.; Syed, A. S.; FohadMabood, H. S., Green synthesis and structural classification of Acacia nilotica mediated-silver doped titanium oxide (Ag/TiO_2) spherical nanoparticles: Assessment of its antimicrobial and anticancer activity, *J. Biol. Sci.* 2019, *26,* 1385–1391.

[54] Weibin, L.; Tamara, L. C.; Andrew, T., Biogenic synthesis of photocatalytically active Ag/TiO_2 and Au/TiO_2 composites, *Harris Chem*., 2012, *14*, 968.

[55] Tri Rahmawati, Teera Butburee, Weradesh Sangkhun, Tuksadon Wutikhun, Jintara Padchasri, Pinit Kidkhunthod, Siripond Phromma, Tippabust Eksangsri, Wiyong Kangwansupamonkon, Pannee Leeladee, Chaweewan Sapcharoenkun; Green synthesis of Ag-TiO_2 nanoparticles using turmeric extract and its enhanced photocatalytic activity under visible light, *Colloids and Surfaces A: Physicochemical and Engineering Aspects,* Volume 665, 2023, 131206, ISSN 0927-775.

[56] Nethravathi, P. C., Udayabhanu, G. Nagaraju et al., TiO_2 and Ag-TiO_2 nanomaterials for enhanced photocatalytic and antioxidant activity: Green synthesis using Cucumis melo juice, Materials Today: Proceedings, *Mater. Today: Proc*, 2021.

[57] Torres-Limiñana, J.; Feregrino-Pérez, A. A.; Vega-González, M.; Escobar-A, L.; Cervantes-Chávez, J. A.; Esquivel, K., Green Synthesis via Eucalyptus globulus L. Extract of Ag-TiO_2 Catalyst: Antimicrobial Activity Evaluation toward Water Disinfection Process, *J. Nanomater*. 2022, *12,* 1944.

[58] Jegadeeswarana, P.; Rajiv, P.; Vanathi, P.; Sivaraj, R.; Rajendran, V., A novel green technology: Synthesis and characterization of Ag/TiO_2 nanocomposites using Padina tetrastromatica (seaweed) extract, *Mater. Lett.*, 2016, *166*, 137-139.

[59] Farnosh, T.; Alireza, B.; Jahan, B. G., Novel and ecofriendly synthesis of Ag-TiO_2-G photocatalyst and investigate the effect of graphene in photodegradation of an organic pollutant in water, *J. Water Environ. Nanotechnol.*, 2019 *4(1),* 31-39.

[60] Pritam, K. D.; Jatin, K.; Amit, K. D.; Soumi S.; Sunita, S.; Swati, S.;, Piyush, K. G.; Beom, S. K.; Green Synthesis of Metallic Nanoparticles: Applications and Limitations, *Catalysts* 2021, *11*, 902.

[61] Anindita, B.; Bharti, M.; Santwana, P.; Nabanita, P.; Jagvir, S. *Multifunctional Hybrid Nanomaterials for Sustainable Agri-food and Ecosystems.* Elsevier, Bhubaneswar, OR, 2020, 639-682.

[62] Mehrdad, K.; Hajar, Q. A.; Iraj S.; Biosynthesis of bimetallic and core–shell nanoparticles: their biomedical applications – a review, *IET Nanobiotechnol.*, 2018, Vol. 12 Iss. 7, pp. 879-887

[63] Sona, A. H.; Parinaz, G.; Bimetallic Nanoparticles with Specific Insight into Nanoremediation, *Importance & Applications of Nanotechnology* (MedDicsebooks) 2020.

[64] Muhammad, S.; Mohsin, S.; Muhammad I.; Nadia, A.; Muhamma, U.; Muhammad, A. A.; Afifa, B. Azadirachta indica leaves extract assisted green synthesis of Ag-TiO_2 for degradation of Methylene blue and Rhodamine B dyes in aqueous medium *Green Process Synth* 2020. vol 39(4).

[65] Slavin, Y. N.; Asnis, J.; Häfeli U. O.; Horacio B.; b Metal nanoparticles: understanding the mechanisms behind antibacterial activity. *J nanobiotechnol* 2017 *15:* 65.

[66] Saee, G.; Rachana, Y.; Balaprasad, A.; Non-antimicrobial and Non-anticancer Properties of ZnO Nanoparticles Biosynthesized Using Different Plant Parts of *Bixa Orellana*, *ACS Omega*, 2022, 7, 7, 1914-1933.

[67] Tooba, M.; Vikas, S.; Rajesh, S. T., Green Synthesis of Bimetallic Nanoparticles and its Applications: A Review, *J. Pharm. Sci. & Res.* 2017, *Vol. 9(2),* 102-110.

[68] Yuqing, C.; Chun, S.; Jie, W.; Gang, X.; Guang, S.; Green Synthesis of Ag-TiO_2 Supported on porous Glass with Enhanced Photocatalytic Performance for Oxidative Desulfurization and Removal of Dyes under Visible Light, *ACS Sustainable Chem. Eng.,* 2018, *6,* 10.

[69] Sibhghatulla, S.; Nazia, N.; Syed Mohd, D. R.; Khurshid, A.; Mohammad, H. B.; Eun, J. L.; Inho C.; Mechanistic Insights into the Antimicrobial Actions of Metallic Nanoparticles and their implications for Multidrug resistance, *Int. J. Mol. Sci.* 2019, 20, 2468.

Chapter 6

Applications of Bimetallic Nanocatalyst in Organic Synthesis: A Comprehensive Overview

Soumyadip Das[1]
and Kaushik Chanda[1,2,*]
[1]Department of Chemistry, School of Advanced Sciences, Vellore Institute of Technology, Vellore, Tamil Nadu, India
[1, 2]Department of Chemistry, Rabindranath Tagore University, Hojai, Assam, India

Abstract

In organic synthesis, metal nanoparticles have emerged as highly adaptable catalysts. Bimetallic nanoparticles have recently captured the attention of chemists as potential catalysts. The introduction of hetero metal-metal connections has altered the surface of the metal catalyst in both electronic and geometrical aspects, leading it to exhibit different activities, selectivities, and stability levels than the comparable monometallic catalyst. In this chapter, the synthesis of newly emerging bimetallic nanocatalysts, such as Cu/Zn, Pd/Au, Co/Pt, Pt/Ni, Pd/Ag, etc., has been discussed. The chapter not only covers the typical catalytic activities of bimetallic catalysts but also delves into the facet-dependent catalytic activities of polyhedral bimetallic nanocrystals, particularly in organic transformations. Furthermore, these bimetallic nanocrystals play a crucial role in sustainable organic transformations.

Keywords: bimetallic nanocatalyst, organic synthesis, green chemistry

[*] Corresponding Author's Email: chandakaushik1@gmail.com.

In: Bimetals
Editors: Pranjal Saikia, Jayanta Kumar Sarmah and Parikshit GogoiName
ISBN: 979-8-89113-496-6

1. Introduction

In organic transformations, catalysts play a crucial role in achieving the complete conversion of starting materials into products under mild reaction conditions. Homogeneous catalysts are often preferred by chemists due to their selectivity, enhanced reactivity, and minimal catalyst loading requirements. Despite their advantages, the use of homogeneous catalysts is frequently hindered by issues such as low catalyst reusability and product contamination with metal. The primary concern with employing homogeneous catalysts lies in the economic loss incurred from the inability to easily recover valuable metal catalysts, especially in industries where reactions are conducted on larger scales [1-2]. To address these limitations associated with homogeneous catalysts, the heterogenization technique was developed. For simple recovery, the homogeneous catalysts are grafted during this heterogenization process. However, there are many drawbacks to using heterogeneous catalysts, including their low reactivity, selectivity, lack of robustness, ability to leach metals from the catalyst, high cost, and laborious design process [3-4].

Therefore, there is a need for a novel catalytic system that combines the easy recoverability of heterogeneous catalysts with the activity characteristic of homogeneous catalysis. As a linker among homogeneous and heterogeneous catalysis, transition metal nanoparticles (sometimes called nanoclusters or colloids) might be an excellent addition as catalyst. Since Nanoparicles characteristically range in magnitude from 1 to 1000 A° (1 A° = 1×10^{-10} m), there is a large availability of surface area, which leads to greater interaction among the starting materials and catalysts and a correspondingly explosive growth in reactivity. Nanoparticles made of transition metals combine the advantages of both systems to produce a novel catalytic activity with high reactivity and selectivity [5].

The simplicity and effectiveness of methods such as precipitation, centrifugation, nanofiltration, and magnetic decantation (particularly for Co, Ni, and Fe), coupled with the potential for product reuse, facilitate the straightforward separation of metal nanoparticles (MNPs).In order to effectively separate and recycle materials, metal nanoparticles have also been stated to simply serve as a strong foundation for a variety of metal complexes as well as organo-catalysts. Particularly, ferrite and magnetite, which are the oxide forms of Fe_2O_3 and Fe_3O_4, respectively, are frequently utilised as a substrate for catalysts. Since the support is magnetic, centrifugation and filtering are not necessary in order to isolate the catalyst by simply utilising outside magnetic field. The emerging area of research to further enhance it is the production of bimetallic nanoparticles. The key characteristic of bimetallic Nanoparticles is the company of two dissimilar kinds of catalytically lively metals that are arranged in close proximity to one another and work together to catalyse the process more efficiently than their monometallic parent metals [6-7].

To develop application of metal nanocatalysis in organic synthesis, our group performed several organic transformations using metal nano catalysts [8-9]. Recently, we have reported the importance of facet-dependent nanocrystals in organic conversion [10]. In this book chapter, we have discussed how to synthesize organic compounds utilising bimetallic nanocrystals that have exceptional catalytic properties. After that, we discussed many organic conversions, such as the C-C coupling, cycloaddition process, hydrogenation of carbonyl, selective hydrogenation, nitroarene reduction, and aryl alkyne hydroboration. Additionally, we discussed the significance of bimetallic nanoparticles in asymmetric synthesis in this book chapter. Our efforts will also result in a greater emphasis on organic synthesis through the use of bimetallic nanocatalyst.

2. Organic Synthesis Using Bimetallic Nanocatalyst

In 2005, Chung and co-workers accomplished the Pauson-Khand-Type reaction using Co/Rh nanocatalyst (Figure 1) The Co_3Rh_2 nanoparticles were use as catalyst for the reaction under 1 atm of CO. The nanocluster $Co_3Rh_2(CO)_{12}$ or $Co_2Rh_2(CO)_{12}$ was synthesized by using a previously reported protocol. Co_2Rh_2 was incorporated into a THF-enyne solution.

There was one atmosphere of carbon monoxide added to the reactor, and it was heated at 130°C for 18 hours. The reaction mixture was allowed to cool to room temperature, then filtered, concentrated, and refined by chromatography using a silica gel column where mobile phase are hexane and diethyl ether (Figure 2) [11].

In 2009, Figuera's group accomplished the asymmetric hydrosilylation utilizing copper-ferrite nanocrystals. Using magnetically separable copper ferrite ($CuFe_2O_4$) nanoparticles (Figure 3), (S)-BINAP as a chiral auxiliary, and PMHS as the stoichiometric reducing agent, room-temperature asymmetric hydrosilylation of ketones were performed and produce chiral alcohols with high yields and high enantiomeric excesses (*ee*) (Figure 4).

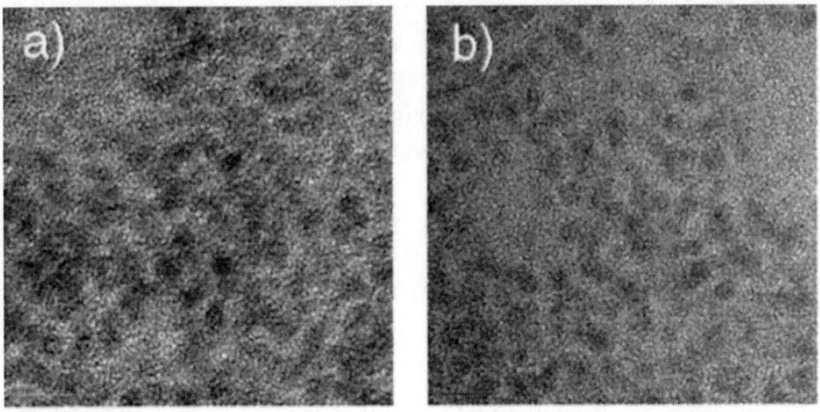

Figure 1. HR-TEM images of (a) Co_2Rh_2 (scale bar 5 nm), (b) Co_3Rh (scale bar 5 nm) Reproduced with permission from ref.11. Copyright 2005 Wiley-VCH.

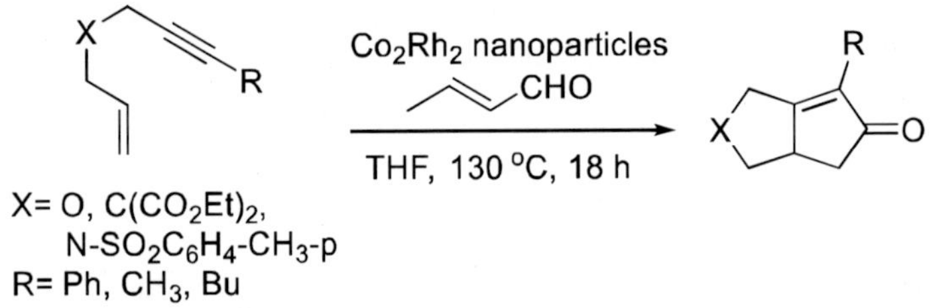

Figure 2. Pauson–Khand reaction catalysed by Rh/Co bimetallic Nanoparticles.

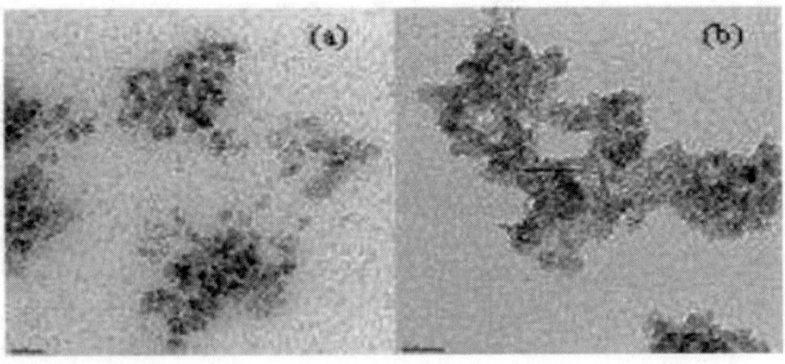

Figure 3. TEM figure of $CuFe_2O_4$ nanoparticles (a) before use, (b) after use. Reproduced with permission from ref.12. Copyright 2009 American Chemical Society.

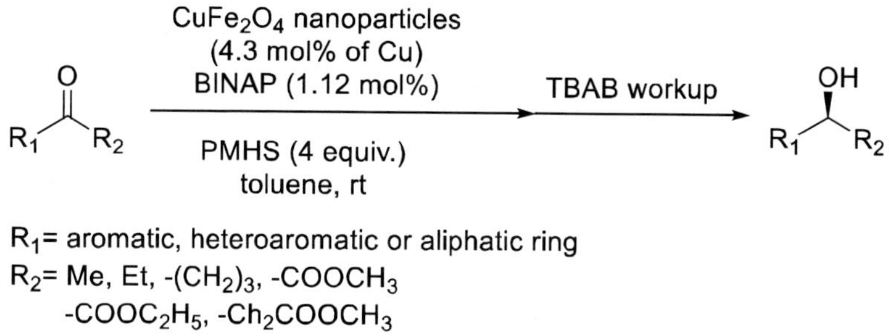

Figure 4. Asymmetric hydrosilylation using copper-ferrite nanocrystals.

$CuFe_2O_4$ nanoparticles were synthesized by co-precipitation protocols as demonstrated previously. One plausible process of hydrosilylation is the production of copper hydride compound via the reduction of ligated copper ferrite nanoparticles by PMHS. A copper alkoxide is created when the copper hydride compound interacts with the carbonyl groups. This copper alkoxide then engages in σ-bond metathesis with the organosilane to create the silyl ether (Figure 5).

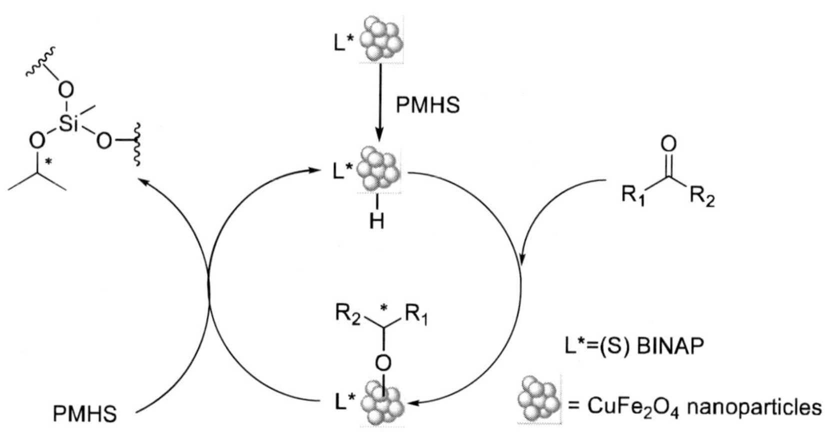

Figure 5. A possible mechanistic pathway for the asymmetric reduction of ketone by Cu/Fe nanoparticles.

Longer reaction times were observed for aromatic ketones with electron-donating groups than for those with electron-withdrawing groups at the para position. The position of the substituents had a significant impact on the enantiomeric excess (ee). Para-substituted ketones produced high ee, while meta- and ortho-substituted ketones resulted in lower values. [12].

In 2009, Yamashita and co-workers created new Fe/Pd supramagnetic nanoparticles modified with chiral BINAP, with Pd(0) forming the particle's shell and Fe occupying the core in its oxidised form (Figure 6). Combination of both metals offered benefits such as Fe (which is easier to separate because its magnetic nature) as well as Pd (which is the catalytically dynamic metal). The catalyst was extensively analysed with different analytical techniques such as SQUID, XAFS, XRD, and TEM. The diameter of nanoparticles was observed to be 56 A°. During the Suzuki-Miyaura cross coupling reaction of 1-bromo-2-methylnaphthylene with 1-naphthylboronic acid in DMF-H_2O solvent at 80°C the catalyst was used. The latent catalytic characteristics of Fe/Pd nanoparticles altered with the use of chiral BINAP. It was discovered that the (S)-BINAP-modified Fe/Pd nanoparticles acts as a chiral nanocatalyst to produce corresponding stereochemical coupled product with more than 99% yield and with a reasonable 48% enantioselectivity. Fe/Pd nanoparticles are believed to be composed of a Fe-rich core and a Pd-rich shell, subsequently Fe atom do not enhance the aforementioned C-C bond formation reaction. No enantioselectivity was observed with OA/OAm stabilized Fe/Pd NPs whereas for (R)-BINAP, (R)-binaphthalene was produced with equal yield and enantioselectivity [13].

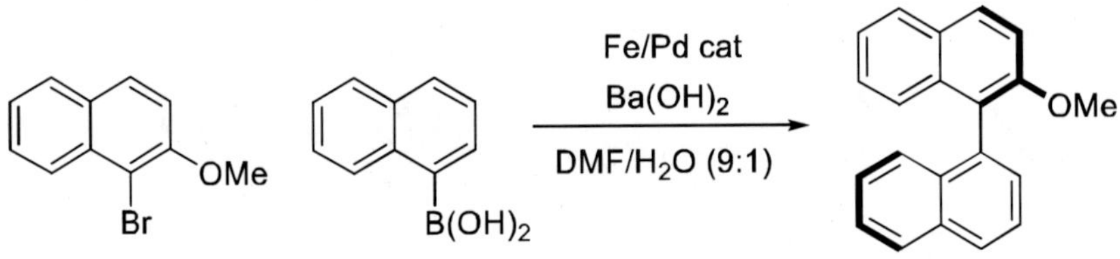

	conv. (%)	ee(%)
Fe/Pd-(R)-BINAP	>99	48
Fe/Pd-(R)-BINAP	>99	48

Figure 6. Fe/Pd nanoparticles with chiral modifications catalyse the Suzuki-Miyaura cross coupling process.

Glorious and group reported the development of Fe_3O_4/Pd nanoparticles (Figure 8) tuned by chiral N-heterocyclic carbenes as well as their utilization of α-arylation. 2-methyl-1-tetralone was reacted with a variety of aryl halides in company of NaO^{t}Bu as a base and produced corresponding arylation product (Figure 9). It was very simple to create the N-heterocyclic carbenes on-site using the base and the appropriate imidazolinium salts. By following the published accounts, Fe_3O_4/Pd nanoparticles were generated using a typical wet impregnation process (Figure 7).

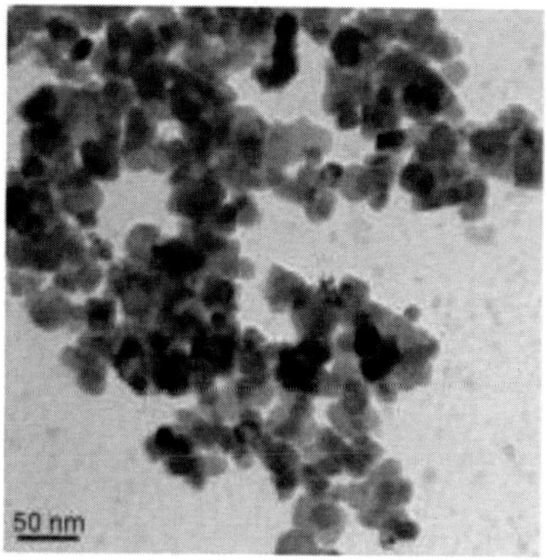

Figure 7. TEM image of Fe_3O_4/Pd NPs after tuned by the ligand. Reproduced with permission from ref.14. Copyright 2010 Wiley-VCH.

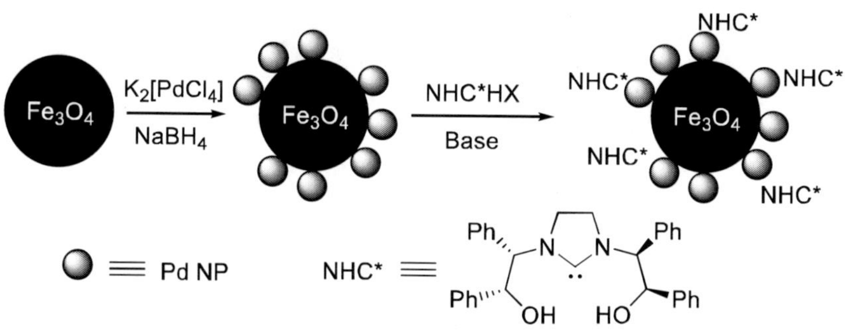

Figure 8. Synthesis of NHC ligand supported Fe_3O_4/Pd nanoparticles.

The scope of the substrates for the enantioselective α-arylation process catalyzed by Fe_3O_4/Pd/L was investigated by studying its reactivity with a variety of aryl halides. Inactivated phenyl halides reacted with 2-methyl-3,4-dihydronaphthalen-1(2*H*)-one, resulted the product in 56% yield and 61% *ee*, but enantioselectivity was slightly lower (45% *ee*) than in the corresponding reaction with iodobenzene. Excellent yields and reasonable to good enantiomeric excess values were obtained from the para- and meta- substituted phenyl bromides. High productivity and discernible enantioselectivities were obtained from the reaction of 2-methyl-3,4-dihydronaphthalen-1(2*H*)-one with phenyl bromides [14].

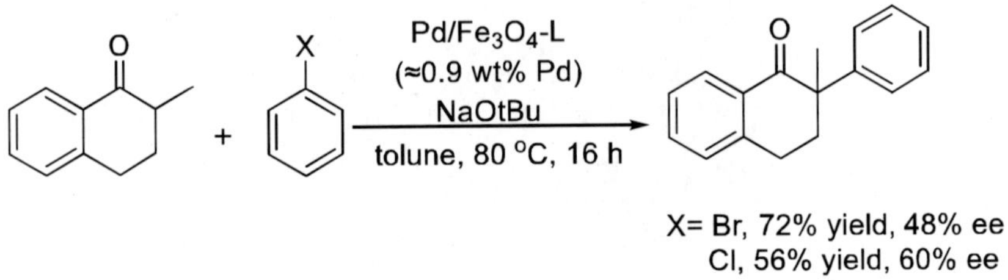

Figure 9. α-arylation of 2-methyl-3,4-dihydronaphthalen-1(2*H*)-one from phenyl halide by using Pd/Fe nanocatalyst.

Figure 10. SEM-EDX Spectrum of Pd-NHC on Fe_3O_4 nanoparticles. Reproduced with permission from ref.15. Copyright 2011 Wiley-VCH.

Glorius et al. recently reported a novel bimetallic catalyst wherein Pd nanoparticles are dispersed on top of Fe nanoparticles and connected to a chiral imidazolium salt through the silane molecule in a two-step process. The preparation of Pd nanoparticles on the surface of NHC-modified MNPs followed a previously reported protocol for platinum nanoparticles supported on Fe_3O_4 nanoparticles (Figure 10).

At first, (3-chloropropyl)triethoxysilane and chiral imidazolium salt was used to modify the Fe_3O_4 NPs. Subsequently, the bimetallic catalyst was obtained by treating chirally altered magnetite nanoparticles with K_2PdCl_4 in hydrazine (Figure 11).

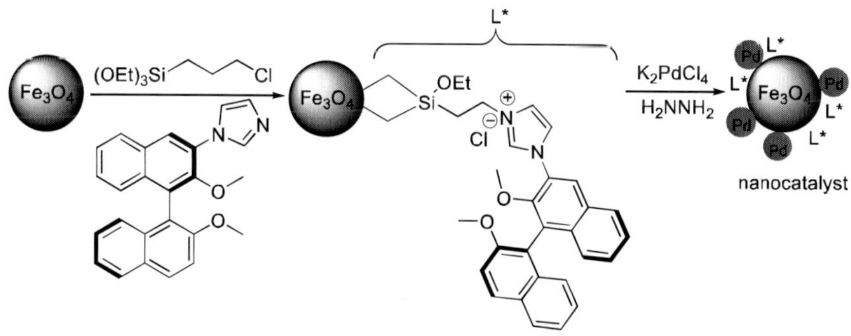

Figure 11. Synthetic pathway of Fe_3O_4/Pd nanoparticles modified by chiral imidazolium.

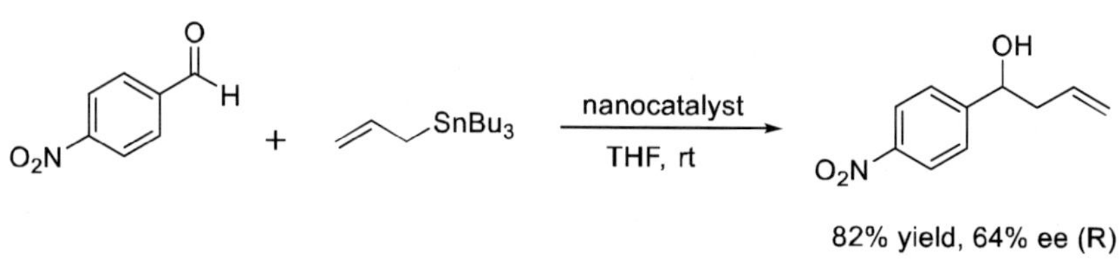

Figure 12. Allylation of 4-nitrobenzaldehyde with Fe_3O_4/Pd nanoparticles.

The modified NPs' catalytic characteristics were assessed, when used as a nanocatalyst in an asymmetric allylation reaction involving 4-nitrobenzaldehyde and allyltributyltin at room temperature in THF. An 82% yield was achieved with a clear enantiomeric excess (*ee*) of up to 64% using the immobilized catalyst, which was shown to be a versatile heterogeneous catalyst (Figure 12) [15].

Kobayashi et al. (2012) described the synthesis of PI/CB Rh/Ag catalyst (Figure 13), an acronym for polymer-integrated with Rh/Ag nanoparticles bound with carbon black (CB). The PI/CB Rh/Ag catalyst was stabilized with a chiral diene ligand for use during high-yield and metal-free 1,4-addition of phenylboronic acids to enones asymmetrically (Figure 14).

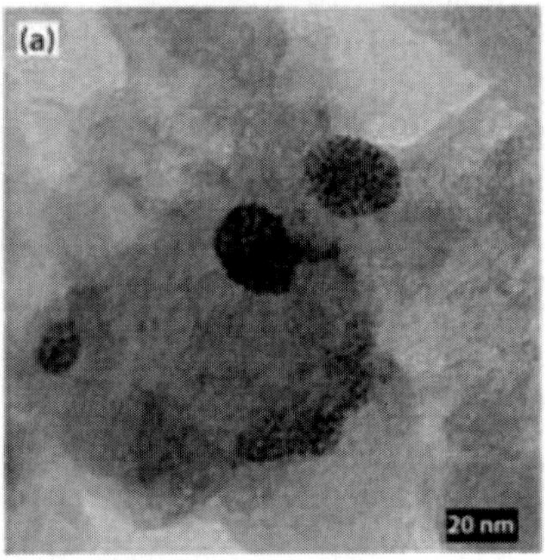

Figure 13. STEM image of PI/CB Rh/Ag catalyst. Reproduced with permission from ref.16. Copyright 2012 American Chemical Society.

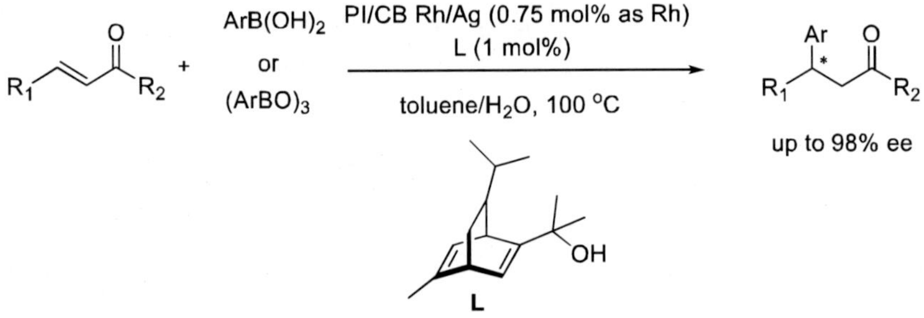

Figure 14. 1, 4-addition reaction by utilizing Rh/Ag catalyst.

More than 99% yield with 98% *ee* was obtained and curiously no Rh leaching was seen. with a bimetallic Rh/Ag with (1:1) ratio as nanocatalyst with chiral diene ligand employed for the asymmetric 1,4-addition reaction of phenylboronic acid and cyclic enone. However, employing the same bimetallic catalyst, leaching of 2.3% or Rh was found in case of acyclic enone. EPS mapping as well as TEM analysis were used to characterize the bimetallic catalyst's structure. Inspecting the NPs, the shape of an alloy was observed. Bimetallic nanocatalyst with Rh and Ag at a ratio of 1:1 bearing diene ligand effectively produce cyclic enones via 1,4-addition reaction and catalysts containing Rh and Ag at a ratio of 1:3 bearing diene ligand effectively generated acyclic enones *via* 1,4-addition reaction of phenylboronic acid and cyclic and acyclic enones [16].

To hydrosilylate prochiral ketones asymmetrically, Shi et al. used chirally modified mesoporous silica KIT-6 supported $CuFe_2O_4$ nanoparticles, where the reduction was performed by using PMHS. Compared to other chiral phosphine ligands, (*S*)-xyl-p-phos ligand performed the best result (Figure 15).

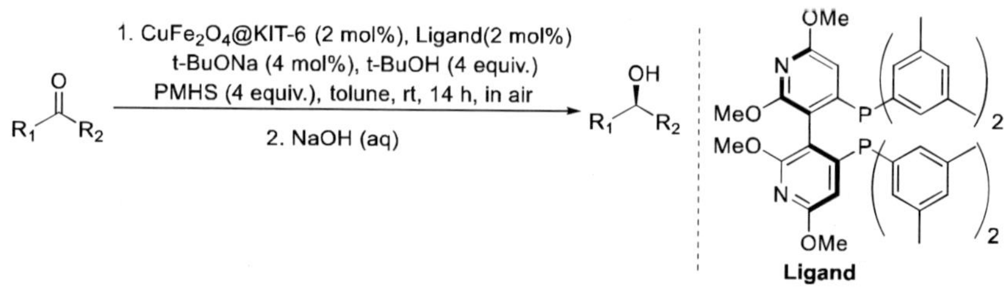

Figure 15. Asymmetric hydrosilylation reaction using mesoporous silica KIT-6 supported $CuFe_2O_4$ nanoparticles.

Although excellent results were generated when the reduction was performed by $PhSiH_3$ during the optimization procedure utilizing the chiral (*S*)-P-Phos as a ligand, PMHS was chosen for the reaction because it is nontoxic, economically practical, and ecologically safe. The reaction was optimized by adding 4 equiv. of *t*-BuOH and 4 mol% of Sodium tert-butoxide, resulting in a 99% yield and 79% *ee* for PMHS. Asymmetric hydrosilylation of phenyl-alkyl prochiral carbonyls was successfully accomplished using the heterogeneous catalysis, which resulted in various beneficial intermediates of biologically active compounds. Quantitative yields with *ee* in the range of 90-96% were obtained by the asymmetric hydrosilylation procedure of halogen-functionalized carbonyls, diaryl carbonyls cycloalkyl carbonyl compounds, and hetero aryl carbonyls compound using the present heterogeneous catalyst (Figure 16).

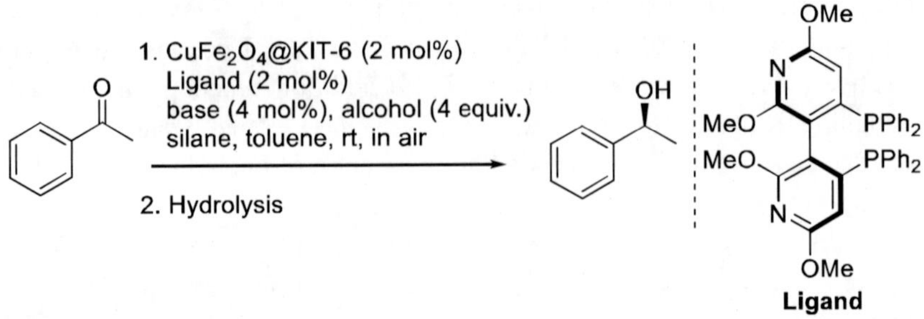

Figure 16. Asymmetric hydrosilylation of acetophenone.

The catalyst demonstrated recyclability for up to four cycles with no significant yield loss and only a minor decrease in enantiomeric excess. Broad as well as small angle XRD patterns of $CuFe_2O_4$ nanoparticles bound with KIT-6 were observed after and before the reaction, validating the consistency and crystal size of the catalyst. SEM and TEM examinations confirmed the results. It was determined that the catalyst's particle size was 10 nm. Improved outcomes were seen because the reaction was accomplished in open air as opposed to nitrogen [17].

In 2013, Kobayashi and group created a heterogeneous multifunctional catalyst incorporate with multi-layer polymers. Polymer beads containing Au/Pd NPs are produced from acryl amides and a styrene-based copolymer (for example, N,N'-methyl enebismethacrylamide, N-tert-butylacrylamide and dimethyl 2,2'-(diazene-1,2-diyl)bis(2-methylpropanoate). It is made so that the nanoparticles are on the outside, while the organic catalyst is on the inside. Dibenzyl malonate as well as cinnamyl alcohol were used as the classical substrates to test the reactivity of the catalysts in a model reaction for the required asymmetric TOP (Figure 17).

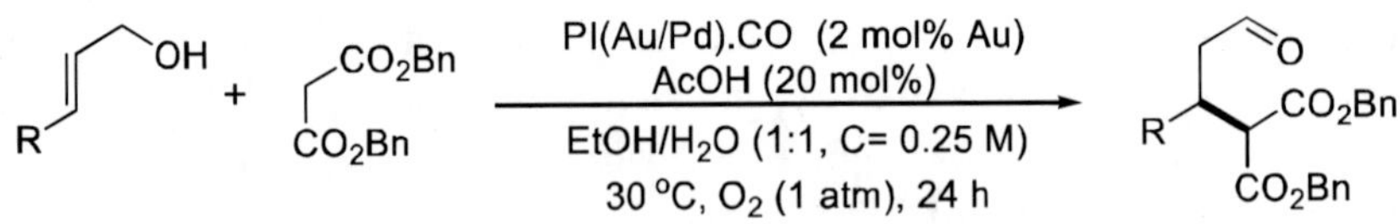

Figure 17. Micheal addition reaction using Au/Pd bimetallic NPs.

To perform the asymmetric 1,4-Michael addition, cinnamyl alcohol would be oxidized aerobically to yield the corresponding trans-cinnamaldehyde utilizing Au-Pd alloy NPs as catalysts. To produce a heterogeneous bifunctional chiral catalyst, Au-Pd alloy NPs are coated with a styrene-based copolymer and immobilized within a polymer-supported OC. By employing this strategy, we were able to avoid potentially harmful interactions between OC and NPs. Usually, the sequence of the two layers containing the distinct catalysts determined the catalytic activity of the composite catalyst for the asymmetric TOP. Providing proof of concept for the efficacy of this innovative method for manufacturing a heterogeneous bifunctional chiral catalyst, this catalyst was found to have a good yield and outstanding enantioselectivity for a wide range of substrates [18].

In 2015, Kobayashi and colleagues presented an enhanced form of PI/CB Rh/Ag nanoparticles (Figure 18), where the chiral modifier is a chiral diene ligand.

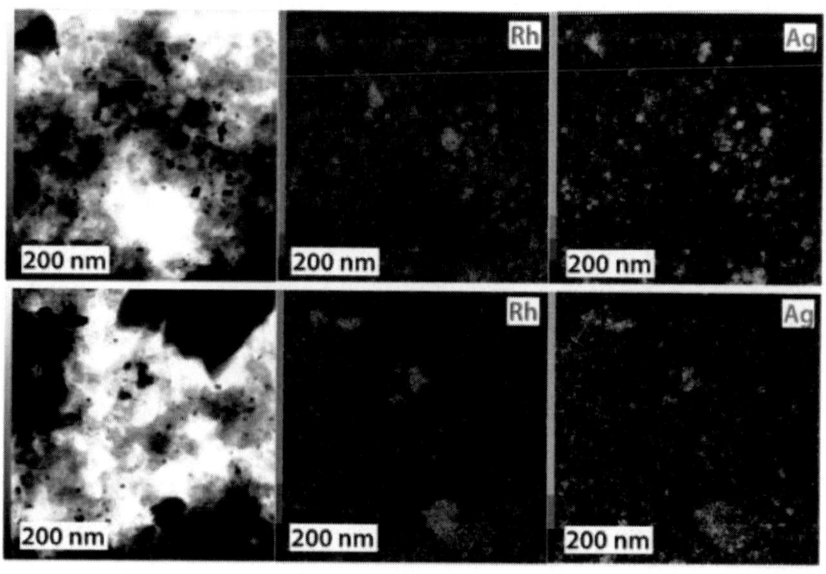

Figure 18. EDS Analysis of PI/CB Rh/Ag nanoparticles. Reproduced with permission from ref.19. Copyright 2015 American Chemical Society.

As an example of C-C bond-formation reaction, the asymmetric 1,4-addition of phenylboronic acids with α, β-unsaturated ketone in presence of the Rh/Ag nanocatalyst was performed (Figure 19). The reactions went without a hitch and produced the desired β-phenylated compounds with moderate to excellent yields, outstanding enantioselectivities, and a variety of substrates. Surprisingly, the performance of the nanoparticle catalysts was outstanding to that of the comparable homogeneous metal catalyst with respect to enantioselectivity, catalytic turnover, and yield. After adding a minimal amount of catalyst, it can be efficiently reused and reused without experiencing a substantial loss of activity for gram-scale synthesis [19].

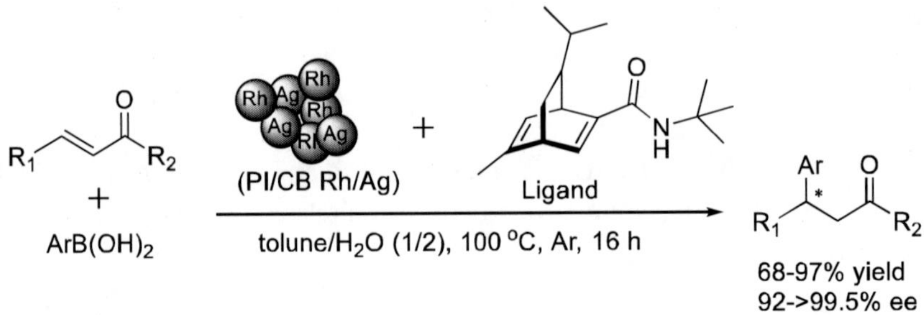

Figure 19. Rh/Ag NPs catalysed 1,4-arylation of α, β-unsaturated carbonyls compound.

The same grump also expanded the use heterogeneous catalyst PI/CB Rh/Ag with chiral diene ligand and for the asymmetric addition of arylboronic acid and α,β-unsaturated amides.

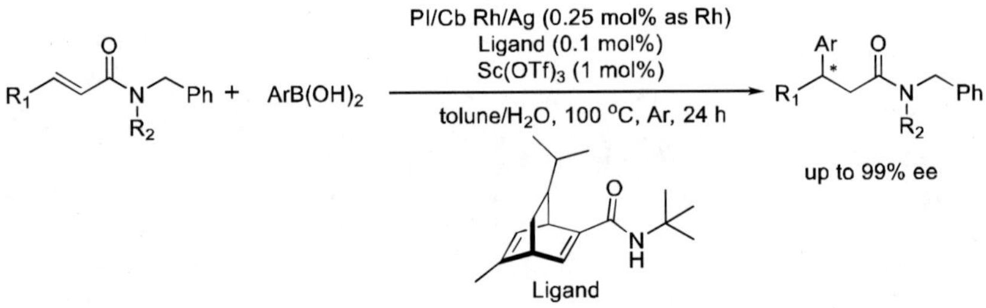

Figure 20. Rh/Ag NPs catalysed asymmetric 1,4-arylation reaction.

$Sc(OTf)_3$, used at 1 mol% as a Lewis acid, was effective at activating the amide substrate, leading to a high yield and excellent *ee*. The pharmaceutically important compound paroxetine was also synthesized using this method as in Figure 20 [20].

Asymmetric arylation was performed using arylboronic acid on aldimines, with polymer-incarcerated (PI/CB Rh/Ag) bimetallic nanoparticles containing chiral diene as ligand. This method produced desired product with excellent yield as well as produce 99% enantiomeric excess as depicted in Figure 21.

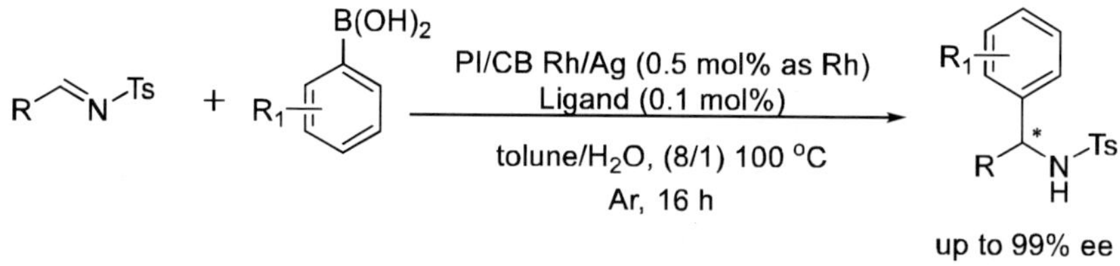

Figure 21. Arylation of aldimines using Rh/Ag bimetallic catalyst.

The reaction took place in water media without significantly hydrolysed imines, producing chiral (diarylmethyl)amines in excellent yields with good enantioselectivities. The maximum turnover number (700) was demonstrated by this catalytic system as heterogeneous catalysts [21].

Suzuki coupling reaction employing Au-Pd core-shell nanocrystals was described by Huang et al. in 2011. Core-shell, truncated octahedra, truncated cubes, cuboctahedra, concave cubes, and octahedra (Figure 22) were created, and tested their catalytic activity. The reaction between $PdCl_2$ and HCl has produced H_2PdCl_4 solution. Only adjusting the concentrations of Au nanocrystals and deionized water allowed for the creation of several core-shell nanocrystal forms using CTAB and ascorbic acid as reducing agent.

Iodobenzene was treated with phenylboronic acid ethanol-water medium at 80°C using Au@Pd nanocrystals and $K_3PO_4.H_2O$ to produce biphenyl with a high yield. Interestingly, concave cubes were bound by the (111) planes, showed greater reactivity in this Suzuki reaction, than the octahedra (Figure 23) [22].

In 2016, Huang et al. used Au-Pd core-shell nanocrystals to demonstrate the 2-amino-5-nirophenol reduction. Tetrahexahedra (THH), cubes, and octahedra with Au-Pd core-shell nanocrystals were prepared for the reduction (Figure 24).

The Au core has been made using the formerly described method. H_2PdCl_4 was employed to create an Au-Pd core-shell THH nanocrystal, and the process also involved the use of CTAC, ascorbic acid, and Au cores. Utilising Au@Pd THH nanoparticles and with the help of H_2O, NH_3BH_3, the 2-amino-5-nitrophenol has been converted to corresponding diaminophenol at room temperature. In this reaction, the Au@Pd THH nanoparticles surrounded by (730) planes showed exceptional activity (Figure 25) [23].

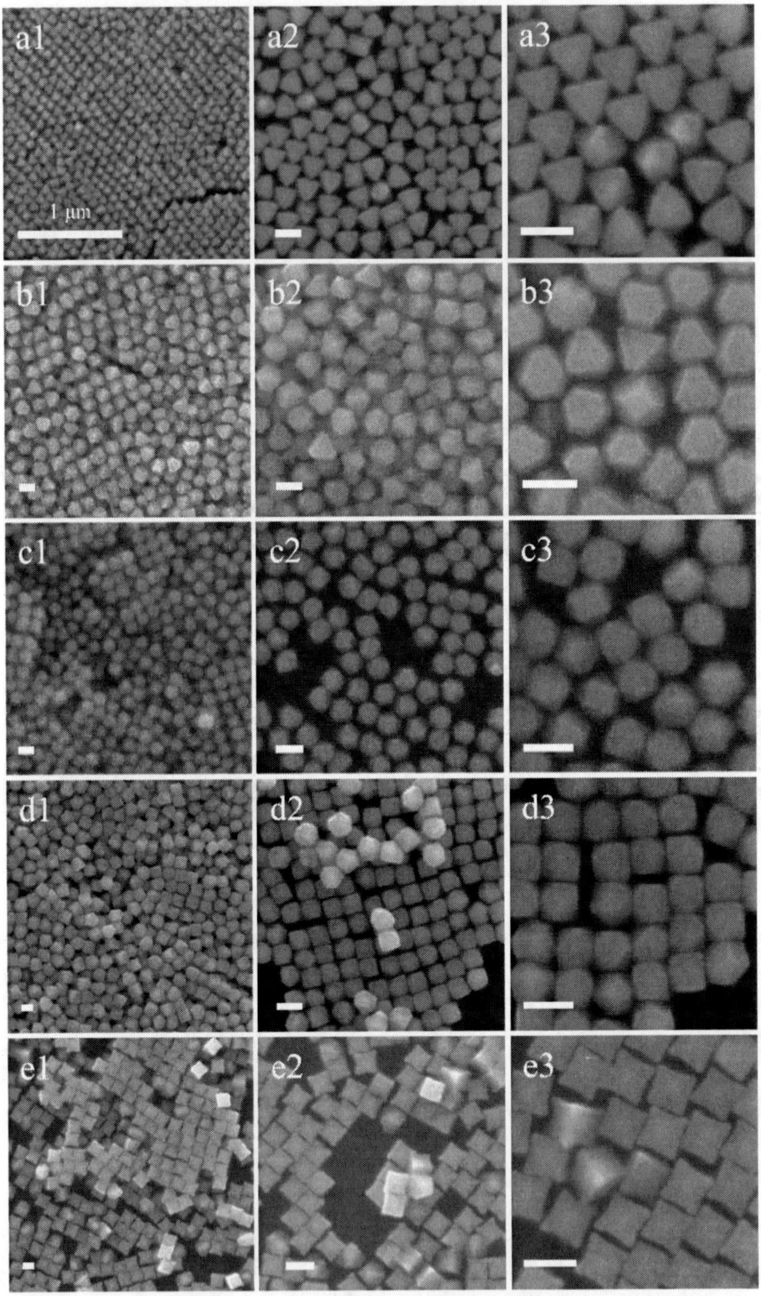

Figure 22. SEM images of Au-Pd core-shell nanocrystals. Reproduced with permission from ref. 22. Copyright 2011 American Chemical Society.

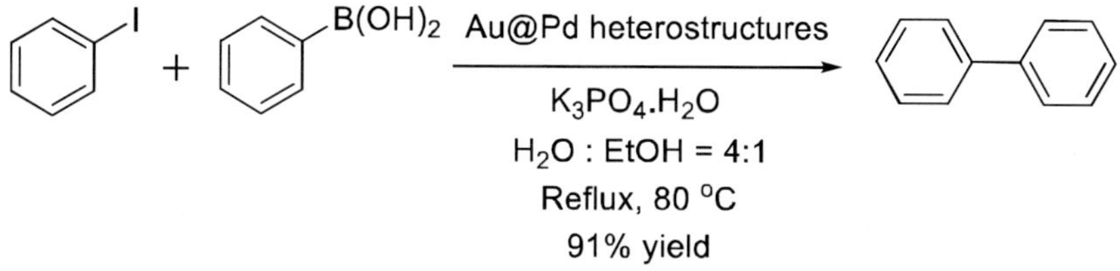

Figure 23. C-C bond forming reaction by utilizng Au@Pd nanocrystals as a catalyst.

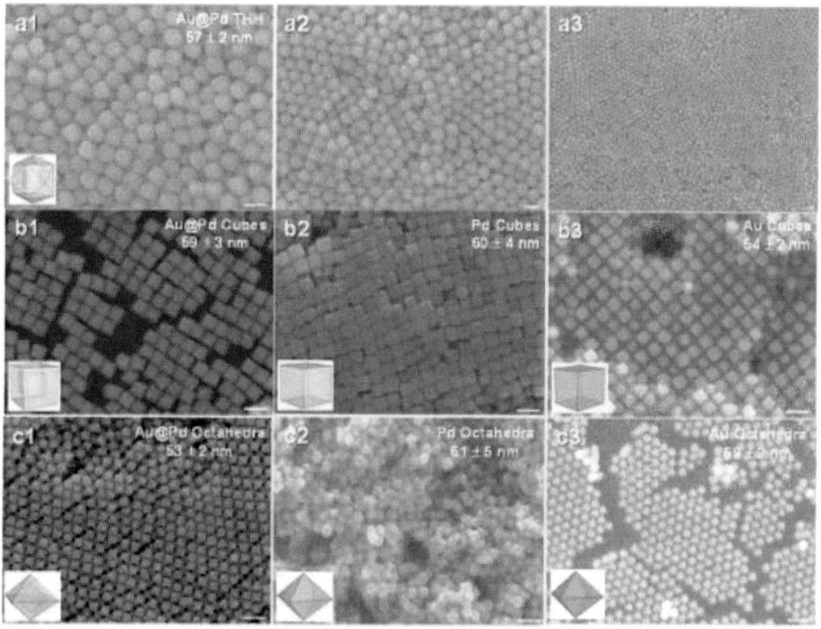

Figure 24. (a) SEM images of THH Au–Pd core-shell nanoparticles. Reproduced with permission from ref. 23. Copyright 2016 Wiley-VCH.

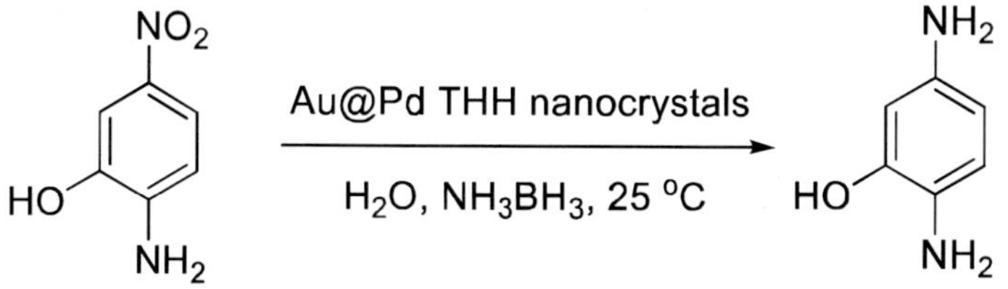

Figure 25. Au@Pd THH NPs catalysed reduction of nitro group.

In 2016, Zhu and co-workers accomplished selectively hydrogenate of α, β-unsaturated aldehydes utilising Pt_xCo_y multicubes and Pt_xCo_y@Co cubes. On reduction of the double bond, the perfect selectivity was displayed by Pt_xCo_y octapod. Selective reduction of the carbonly bond of α, β-unsaturated carbonyl compound was demonstrated by Pt_xCo_y multicubes. Pt_2Co_1 multicubes demonstrated superior catalytic activity in the, α, β-unsaturated aldehyde hydrogenation process compared to standard Pt/C catalyst and Pt nanocubes. Pt_2Co_1 is surrounded by several high-index faces, of which facet (130) produced the best catalytic activity [24].

In 2016, Zhu and his coworkers successfully utilized Pt_xCo_y nanoparticles with curved faces for the hydrogenation of cinnamaldehyde. In the reduction of cinnamaldehyde, the $Pt_{11}Co$ concave cubes bound by (710) planes exhibited better selectivity and greater activity than $Pt_{11}Co$ cubes bound by (100) planes. [25].

In 2017, Huang and his team demonstrated the synthesis of 1,4-disubstituted triazoles using Au/Cu core-shell nanocubes. Following a previously described method, the Au-Cu nanocrystals were prepared $CuCl_2$ or copper acetate was used to create these Au-Cu nanocrystals, such as cubes and octahedra, which were then heated to 100°C for 45 minutes while also containing ascorbic acid, hexadecyl amine, and Au cores. At 50°C and in the company of Au@Cu nanocubes, benzyl azide as well as ethynyl benzene interacted in a water medium to produce a good yield of 1,4-disubstituted triazole. Au@Cu cubes having a facet (100) and octahedra having a facet (111), respectively, showed greater catalytic activity than the former one (Figure 26) [26].

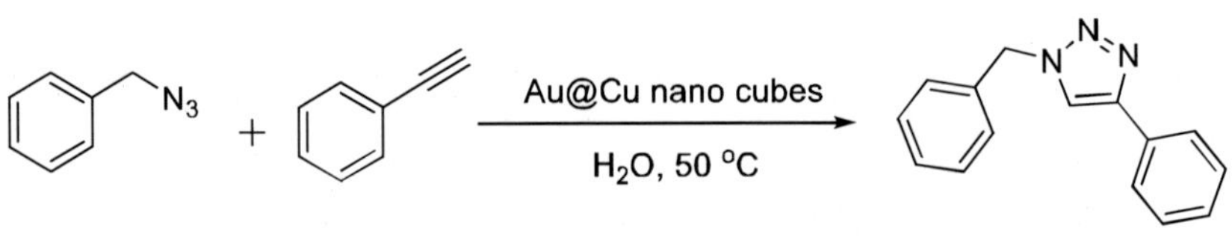

Figure 26. Au@Cu nanocubes catalysed synthesis of 1,4-functionalized triazoles.

In 2019, Pal and his coworkers employed TiO_2 nanosheets for the selective hydrogenation of quinoline, utilizing a co-catalyst consisting of Cu(core)@Zn(shell) under visible light. $CuCl_2$, $ZnCl_2$, and aqueous gelati solution were first combined to create a homogenous solution. Then, using a reduction technique with glucose as the reducing agent, Cu@Zn core-shell nanocrystals were created. The Cu/Zn-TiO_2 nano materials were made using the formerly described methodology, and the TiO_2 nanosheets surrounded with (100) planes were created using hydrofluoric acid, Titanium butoxide, and a stainless-steel autoclave with Teflon lining. Using H_2 and a Cu@Zn-TiO_2 nanocatalyst, quinoline was reduced to 1,2,3,4-terahydoquinoline in presence of visible light. After Cu@Zn was impregnated on TiO_2, the (001) facet-enclosed TiO_2 nanosheet demonstrated exceptional catalytic activity (Figure 27) [27].

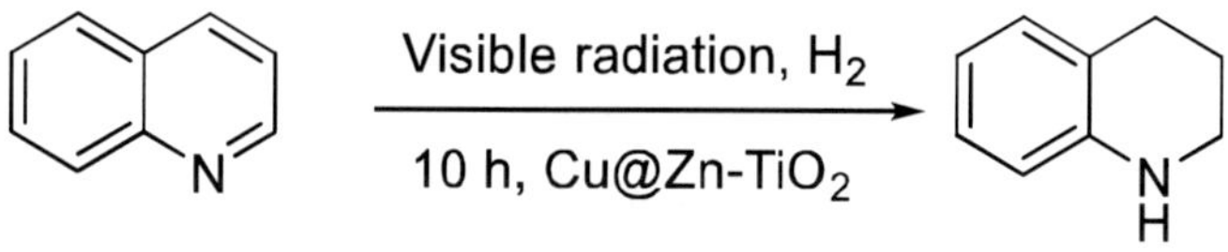

Figure 27. Selective semi-hydrogenation of quinolone moiety using Cu@Zn-TiO_2 nanocatalyst.

Fang et al. reported a straightforward colloidal synthesis approach for preparing CuNi nano-octahedra and nanocubes. CuNi nanocrystals was formed from a group of synergetic capping ligands controlled and in a hot solution environment, giving them distinctive crystallographic facets at the ends. In particular, the growth of (111)-bounded CuNi nano-octahedra is controlled by thermodynamics, whereas the kinetic capping of chloride involved for the production of (100)-terminated CuNi nanocubes. CuNi nanocubes as well as nano-octahedra (with facet dependence properties) both displayed outstanding activity in the hydrogenation of nitro group with $NaBH_4$. In comparison to CuNi nano-octahedra, CuNi nanocubes demonstrated better catalytic activity, as (100) facets absorb more nitroarene and hydrogen than (111) facet of nano-octahedra [28].

Conclusion

It is relatively straightforward to create bimetallic catalysts by combining corresponding metal precursors, combining two metals with catalytic activity into a single molecule with the assistance of chiral functionalities. For easy recovery, catalysts are often immobilized on a solid support. Bimetallic catalysts commonly show enhanced activity, leading to alterations in the chemical characteristics of the catalyst and expanding the potential for organic synthesis. It can be reused repeatedly without a significant loss of catalytic activity. Leaching of the metal was minimal or non-existent in almost all instances of bimetallic catalyst. By combining their effects, the two metals in the catalyst fine-tune each other's catalytic capabilities and catalyse the reaction more successfully. In addition to simplifying the separation process, the incorporation of Fe metals in the bimetallic nanocatalyst also enhances the catalyst's cost-effectiveness. These heterogeneous catalysts frequently produce better results than their homogeneous counterparts, and in some situations, special selectivity was also seen, suggesting that they should be referred to as different active species rather than "immobilised catalysts." Excellent yields and good enantiomeric excess were obtained in majority of cases, and it was discovered that the *ee* was primarily dependent on the chiral modifier employed. These heterogeneous catalysts often outperform their homogeneous counterparts, and in some cases, exclusive selectivity has been observed. This has led some researchers to argue that they should not be labeled as "immobilized catalysts" but rather as distinct active species.

Acknowledgments

The authors acknowledge the Chancellor and Vice Chancellor of VIT University for delivering the opportunity to carry out this study. KC thanks CSIR-GOI for funding through grant no 01(2913)/17/EMR-II.

References

[1] Baleizão, C.; Garcia, H. Chiral Salen Complexes: An Overview to Recoverable and Reusable Homogeneous and Heterogeneous Catalysts. *Chem. Rev.* 2006, *106* (9), 3987–4043.

[2] Corma, A.; Garcia, H. Silica-Bound Homogenous Catalysts as Recoverable and Reusable Catalysts in Organic Synthesis. *Adv. Synth. Catal.* 2006, *348* (12–13), 1391–1412.

[3] Alaerts, L.; Wahlen, J.; Jacobs, P. A.; De Vos, D. E. Recent Progress in the Immobilization of Catalysts for Selective Oxidation in the Liquid Phase. *Chem. Commun.* 2008, No. 15, 1727–1737.

[4] Zhao, D.; Ding, K. Recent Advances in Asymmetric Catalysis in Flow. *ACS Catal.* 2013, *3* (5), 928–944.

[5] Yan, N.; Xiao, C.; Kou, Y. Transition Metal Nanoparticle Catalysis in Green Solvents. *Coord. Chem. Rev.* 2010, *254* (9), 1179–1218.

[6] Favier, I.; Gómez, M.; Muller, G.; Axet, M. R.; Castillón, S.; Claver, C.; Jansat, S.; Chaudret, B.; Philippot, K. Palladium Catalytic Species Containing Chiral

Phosphites: Towards a Discrimination between Molecular and Colloidal Catalysts. *Adv. Synth. Catal.* 2007, *349* (16), 2459–2469.

[7] Sasaki, K.; Naohara, H.; Cai, Y.; Choi, Y. M.; Liu, P.; Vukmirovic, M. B.; Wang, J. X.; Adzic, R. R. Core-Protected Platinum Monolayer Shell High-Stability Electrocatalysts for Fuel-Cell Cathodes. *Angew. Chemie Int. Ed.* 2010, *49* (46), 8602–8607.

[8] Panchangam, R. L.; Manickam, V.; Chanda, K. Assembly of Fully Substituted 2H-Indazoles Catalyzed by Cu2O Rhombic Dodecahedra and Evaluation of Anticancer Activity. *ChemMedChem* 2019, *14* (2), 262–272.

[9] Padmaja, R. D.; Rej, S.; Chanda, K. Environmentally Friendly, Microwave-Assisted Synthesis of 5-Substituted 1 H-Tetrazoles by Recyclable CuO Nanoparticles via (3+2) Cycloaddition of Nitriles and NaN_3. *Chinese J. Catal.* 2017, *38* (11), 1918–1924.

[10] Das, S.; Chanda, K. The Role of Nanocrystal Facets in Sustainable Organic Synthesis. *ChemNanoMat* 2022, *8* (11), e202200375.

[11] Park, K. H.; Chung, Y. K. Immobilized Co/Rh Heterobimetallic Nanoparticle-Catalyzed Pauson–Khand-Type Reaction. *Adv. Synth. Catal.* 2005, *347* (6), 854–866.

[12] Kantam, M. L.; Yadav, J.; Laha, S.; Srinivas, P.; Sreedhar, B.; Figueras, F. Asymmetric Hydrosilylation of Ketones Catalyzed by Magnetically Recoverable and Reusable Copper Ferrite Nanoparticles. *J. Org. Chem.* 2009, *74* (12), 4608–4611.

[13] Mori, K.; Kondo, Y.; Yamashita, H. Synthesis and Characterization of FePd Magnetic Nanoparticles Modified with Chiral BINAP Ligand as a Recoverable Catalyst Vehicle for the Asymmetric Coupling Reaction. *Phys. Chem. Chem. Phys.* 2009, *11* (39), 8949–8954.

[14] Ranganath, K. V. S.; Kloesges, J.; Schäfer, A. H.; Glorius, F. Asymmetric Nanocatalysis: N-Heterocyclic Carbenes as Chiral Modifiers of Fe_3O_4/Pd Nanoparticles. *Angew. Chemie - Int. Ed.* 2010, *49* (42), 7786–7789.

[15] Yasukawa, T.; Miyamura, H.; Kobayashi, S. Polymer-Incarcerated Chiral Rh/Ag Nanoparticles for Asymmetric 1,4-Addition Reactions of Arylboronic Acids to Enones: Remarkable Effects of Bimetallic Structure on Activity and Metal Leaching. *J. Am. Chem. Soc.* 2012, *134* (41), 16963–16966.

[16] Yasukawa, T.; Miyamura, H.; Kobayashi, S. Polymer-Incarcerated Chiral Rh/Ag Nanoparticles for Asymmetric 1,4-Addition Reactions of Arylboronic Acids to Enones: Remarkable Effects of Bimetallic Structure on Activity and Metal Leaching. *J. Am. Chem. Soc.* 2012, *134* (41), 16963–16966.

[17] Li, M.; Li, B.; Xia, H. F.; Ye, D.; Wu, J.; Shi, Y. Mesoporous Silica KIT-6 Supported Superparamagnetic CuFe2O 4 Nanoparticles for Catalytic Asymmetric Hydrosilylation of Ketones in Air. *Green Chem.* 2014, *16* (5), 2680–2688.

[18] Miyamura, H.; Choo, G. C. Y.; Yasukawa, T.; Yoo, W. J.; Kobayashi, S. A Heterogeneous Layered Bifunctional Catalyst for the Integration of Aerobic Oxidation and Asymmetric C-C Bond Formation. *Chem. Commun.* 2013, *49* (85), 9917–9919.

[19] Yasukawa, T.; Suzuki, A.; Miyamura, H.; Nishino, K.; Kobayashi, S. Chiral Metal Nanoparticle Systems as Heterogeneous Catalysts beyond Homogeneous Metal Complex Catalysts for Asymmetric Addition of Arylboronic Acids to α,β-Unsaturated Carbonyl Compounds. *J. Am. Chem. Soc.* 2015, *137* (20), 6616–6623.

[20] Yasukawa, T.; Saito, Y.; Miyamura, H.; Kobayashi, S. Chiral Nanoparticles/Lewis Acids as Cooperative Catalysts for Asymmetric 1,4-Addition of Arylboronic Acids to α,β-Unsaturated Amides. *Angew. Chemie - Int. Ed.* 2016, 55(28), 8058-8061.

[21] Yasukawa, T.; Kuremoto, T.; Miyamura, H.; Kobayashi, S. Asymmetric Arylation of Imines Catalyzed by Heterogeneous Chiral Rhodium Nanoparticles. *Org. Lett.* 2016, *18* (11), 2716–2718.

[22] Yang, C.-W.; Chanda, K.; Lin, P.-H.; Wang, Y.-N.; Liao, C.-W.; Huang, M. H. Fabrication of Au–Pd Core–Shell Heterostructures with Systematic Shape Evolution Using Octahedral Nanocrystal Cores and Their Catalytic Activity. *J. Am. Chem. Soc.* 2011, *133* (49), 19993–20000.

[23] Rej, S.; Hsia, C. F.; Chen, T. Y.; Lin, F. C.; Huang J. S.; Huang, M. H. Facet-Dependent and Light-Assisted Efficient Hydrogen Evolution from Ammonia Borane Using Gold-Palladium Core-Shell Nanocatalysts *Angew. Chemie - Int. Ed.* 2016, 55(25), 7222-7226.

[24] Huang, S.; Yang, N.; Huo, M.; Sun, Y.; Zhu, Y. The Shape Evolution from PtxCoy@Co Cubes to PtxCoy Multicubes for Selective Hydrogenation of α,β-Unsaturated Aldehyde. *Nanoscale* 2016, *8* (12), 6451–6455.

[25] Gu, Y.; Zhao, Y.; Wu, P.; Yang, B.; yang, N.; Zhu, Y. Bimetallic PtxCoy Nanoparticles with Curved Faces for Highly Efficient Hydrogenation of Cinnamaldehyde. *Nanoscale* 2016, *8* (21), 10896–10901.

[26] Madasu, M.; Hsia, C.-F.; Huang, M. H. Au–Cu Core–Shell Nanocube-Catalyzed Click Reactions for Efficient Synthesis of Diverse Triazoles. *Nanoscale* 2017, *9* (21), 6970–6974.

[27] Bathla, A.; Pal, B. Bimetallic Cu(Core)@Zn(Shell) Co-Catalyst Impregnated TiO_2 Nanosheets (001 Faceted) for the Selective Hydrogenation of Quinoline under Visible Light Irradiation. *J. Ind. Eng. Chem.* 2019, *79*, 314–325.

[28] Li, C.; Luan, Y.; Zhao, B.; Kumbhar, A.; Chen, X.; Collins, D.; Zhou, G.; Fang, J. Facet-Dependent Catalysis of CuNi Nanocatalysts toward 4-Nitrophenol Reduction Reaction. *MRS Adv.* 2020, *5* (27), 1491–1496.

Chapter 7

Recent Advances in Green Synthesized rGO/TiO_2-Based Bimetallic Heterojunctions for Organic Pollutant Degradation

Nandini Priyam Rajkumari[1]
Kalyani Adhikary[2]
Sahima Siddika[2]
and Pallabi Goswami[$,*]
[1]Department of Chemistry, Gauhati University, Guwahati, Assam, India
[2]Department of Chemistry, Nowgong College (Autonomous), Nagaon, Assam, India

Abstract

Green synthesis of reduced graphene oxide/titanium dioxide (rGO/TiO_2)-based bimetallic heterojunctions for environmental remediation has gained significant attention in recent years. These advanced materials offer enhanced photocatalytic properties and hold great promise for the degradation of various pollutants in the water and air. Green synthesis techniques have become a necessity due to numerous recent advancements in this field. Traditional synthesis routes for rGO/TiO_2 heterojunctions often involve the use of harsh chemicals and high temperatures, which can be environmentally detrimental. In contrast, recent research has focused on environmentally friendly and sustainable synthesis techniques. The incorporation of bimetallic nanoparticles into the rGO/TiO_2 heterojunctions has been found to significantly enhance the photocatalytic activity. The introduction of noble metal nanoparticles (e.g., gold, silver, platinum) or non-noble metal nanoparticles (e.g.,

* Corresponding Author's Email: gpallabi33@gmail.com.

In: Bimetals
Editors: Pranjal Saikia, Jayanta Kumar Sarmah and Parikshit GogoiName
ISBN: 979-8-89113-496-6

copper, nickel, iron) onto the heterojunctions can enhance the light absorption and promote the separation of photogenerated electron-hole pairs. This leads to improved efficiency in pollutant degradation and reduces the recombination rate of charge carriers, also by tailoring the properties of rGO/TiO_2 bimetallic heterojunctions lead to selectively remove specific pollutants. The construction of heterostructured frameworks within the rGO/TiO_2 heterojunctions has been assessed in this chapter as a way to further improve the photocatalytic efficiency. These heterostructure architectures provide additional active sites and pathways for pollutant degradation, leading to improved environmental remediation performance. Therefore, this chapter gives an overview of green synthesized rGO/TiO_2-based bimetallic heterojunctions that have shown great potential for environmental remediation.

Keywords: green synthesised rGO, titanium dioxide, bimetallic, heterojunctions

1. Introduction

Reduced graphene oxide/titanium dioxide (rGO/TiO_2)-based bimetallic heterojunctions are hybrid materials composed of reduced graphene oxide (rGO) and titanium dioxide (TiO_2) nanoparticles, with the addition of one or more metallic components [1]. These heterojunctions have attracted significant attention in pollutant degradation and other applications due to their unique properties and synergistic effects [2]. The combination of reduced graphene oxide (rGO) and titanium dioxide (TiO_2) in a heterojunction structure offers several advantages [3]. Titanium dioxide (TiO_2) is a well-known photocatalyst with excellent light absorption and electron-hole separation capabilities [4]. By incorporating reduced graphene oxide (rGO), which acts as an electron conductor and enhances charge transfer, the photocatalytic activity of titanium dioxide (TiO_2) can be further improved (Figure 1)[5]. The bimetallic component added to the heterojunction can also promote charge separation and reduce recombination, leading to enhanced photocatalytic efficiency [6]. Titanium dioxide (TiO_2) predominantly absorbs UV light, limiting its photocatalytic activity to the UV region of the electromagnetic spectrum [1-6]. However, reduced graphene oxide (rGO) exhibits broadband light absorption, including visible light. Integrating reduced graphene oxide (rGO) with titanium dioxide (TiO_2) expands the light absorption range, allowing for more efficient utilization of solar energy. The combination of reduced graphene oxide (rGO) and titanium dioxide (TiO_2) in a heterojunction structure creates synergistic effects [7]. Reduced graphene oxide (rGO) provides a large surface area, promotes adsorption of pollutants, and facilitates electron transfer, while titanium dioxide (TiO_2) offers strong oxidative properties and catalytic activity [8]. The synergy between these materials leads to improved pollutant degradation and remediation efficiency. By incorporating bimetallic components into the heterojunction, selectivity towards specific pollutants can be achieved [9]. Different bimetallic combinations have been explored to target the degradation of organic dyes, heavy metals, antibiotics, pesticides, and other pollutants. The choice of bimetallic components depends on the nature of the pollutants and their chemical interactions with the heterojunction. The combination of reduced graphene oxide (rGO) and titanium dioxide (TiO_2) enhances the stability and recyclability of the heterojunction [10]. Reduced graphene oxide (rGO) acts as a stabilizing agent, preventing agglomeration and improving the structural integrity of titanium dioxide (TiO_2) nanoparticles. This leads to increased durability and longevity of the heterojunction, enabling multiple cycles of pollutant degradation without significant loss of activity [7-10].

Different semiconductor photocatalysts used so far i.e., TiO_2, Fe_2O_3, CuS, CdS, Zn^O, Ni^O, ZnS, CdSe, $ZnFeO_4$ Ag_3PO_4, MnO_2, Cu_2O $BiWO_6$ etc [11]. Most researched photocatalyst is however, titanium dioxide (TiO_2). Reasons being its long term thermodynamic stability, strong oxidizing power, low cost and relative nontoxicity [12]. But titanium dioxide (TiO_2) cannot be used as lone component [13]. When it is used as single component its efficiency is found to be very low. Also the range of wavelength absorb is very less [14-17]. Titanium dioxide (TiO_2) has a band gap of only 3.2 eV, which allows it to be active in the UV range only [18, 19]. A 3.2 eV electronic band gap makes titanium dioxide (TiO_2) unfeasible for use under natural sunlight below a wavelength of ~380nm. Apart from this, the quantum yields as well as the efficiency for photocatalytic reactions also decrease considerably because of the very high rate of electron and hole recombination. A number of schemes have been employed to solve these problems, such as doping with impurities in the form of metals and nonmetals (substitutional or/and interstitial), [20] modification of the parent photocatalyst in terms of surface properties or nanocomposite formation with other supplementary materials (e.g., heterojunctions), pairing with different semiconductors of varying band gaps [21] and packing noble metals.

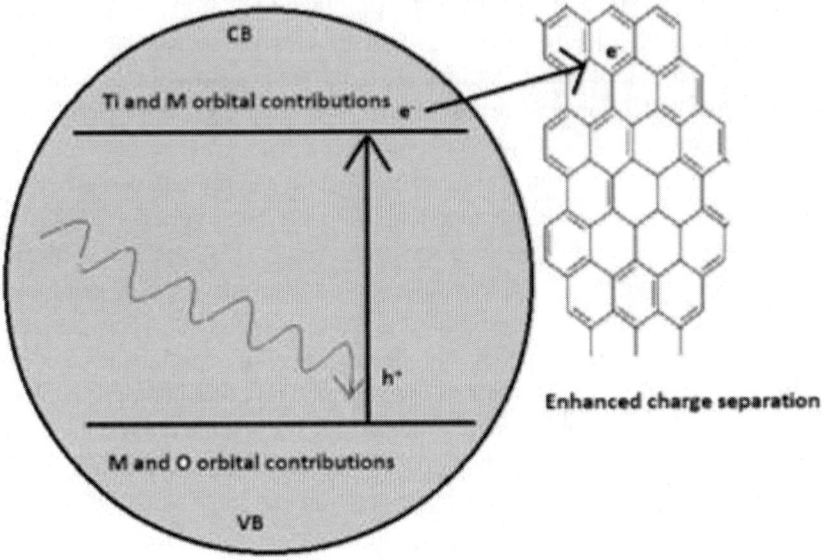

Figure 1. Production of electron and hole in the photocatalyst and migration of excited electron to reduced graphene oxide.

In particular, the amalgamation of titanium dioxide (TiO_2) nanomaterials with carbon-based nano raw materials such as graphene, graphite oxide, and carbon nanotubes has been proposed as a convenient method for enhancing the photocatalytic activity. Currently, graphene oxide (GO) and reduced graphene oxide (rGO) are utilized as the most effective support systems for enhancing photocatalytic activity as they possess unique physical as well as chemical properties [22], E.g., superior mechanical strength, high conductivity, extended specific surface area, etc. The overall effect of utilizing graphene oxide (GO) and reduced graphene oxide (rGO) to synthesize a GO/rGO/TiO_2 nanocomposite is to enhance the photoreactivity of titanium dioxide (TiO_2) by reducing the excited electron and hole recombination, thereby increasing the light absorption range. The fusion of titanium dioxide (TiO_2) with reduced graphene oxide (rGO) is one of the best methods to overcome the drawbacks of titanium dioxide (TiO_2) [23]. Reduced graphene oxide (rGO) has considerable utility in a wide range of areas. The reduced graphene oxide (rGO)-dependent nanocomposites find uses as nanosensors for various biomolecules, photocatalysts for degradation of pollutants, methanol oxidation, lithium-ion batteries, etc. Graphene oxide (GO), or reduced graphene oxide (rGO) coupled with TiO_2, has been found to show significant electron-withdrawing as well as electron-storing ability [24]. Therefore, the nanocomposites of GO/TiO_2 or rGO/TiO_2 have received considerable attention from the scientific community in a large number of photofunctional applications such as purification of water and air photocatalytically, in water splitting for production of hydrogen as well as oxygen, in solar cells, and in bacterial inactivation [25].

The function of graphite, graphene oxide, and reduced graphene oxide is proposed as follows. Under the irradiation of UV light of suitable frequency in titania, electrons excited from the conduction band (CB) migrate to the graphene oxide (GO) or reduced graphene oxide (rGO) matrix, which has a more positive Fermi level and positively hinders the recombination of electrons and holes [26]. However, the reverse of this process takes place upon excitation by visible light, where electrons transfer from the photoexcited state of graphene to conduction band (CB) of titania.

When oxidizing agents react with graphite, graphene oxide is produced, which is a byproduct of this oxidation formed efficiently as oxidizing agents increase the interpolation interval between the layers of graphite [27]. The completely oxidized compound can then disperse into the base material like water and graphene oxide is formed. Generally, carbon-based nanomaterials have been elaborately studied for various catalytic applications either as the catalyst by itself or by serving as an aiding matrix. Graphene is used to monitor and then enhance the electronic or photonic properties of the original catalyst of a particular electrochemical or photochemical reaction [28]. Unlike photonic and electronic applications, which are possible only with high-purity sp^2-hybridized carbon-containing graphene that can only be produced by delicate approaches, it is expected that for solving our day-to-day energy or environment-related problems, we require methods that are cost-effective and produce large quantities easily and also help us to produce flawed single-layer or few-layer graphene with surface functionalization [29]. Also, we need methods that can be quickly operated through the well-known Hummers' process. Instead of giving a single layer, graphene Hummer's method gives graphene oxide (GO) [30]. The resulting graphene oxide (GO) contains plenty of hydrophilic groups, namely, carboxyl, hydroxyl, and epoxy groups, which are bonded with carbon. In Hummer's process, the synthesized graphene oxide (GO) has significantly distorted sp^2-hybridized carbon with in-plane bonding that acts as scattering heads that adequately alter the surface conductivity and optoelectronic properties of graphene oxide (GO), which otherwise possess low transparency, charge carrier mobility, and high resistivity [31]. Graphene oxide, upon reduction and rearrangement of the sp^2 aromatic structures, leads to the formation of reduced graphene oxide (rGO), which exhibits proficiency in charge shuttling and provides excellent catalytic sites. Reduction strategies involving various reducing agents, such as photochemical, hydrothermal, and chemical reagents like hydrazine, have been found to be useful for the preparation of reduced graphene oxide (rGO) (Figure 2) [32].

Graphene is essentially a single atomic layer of graphite, consisting of sp^2 bonded carbon atoms arranged in a hexagonal or honeycomb lattice [33]. Graphene is fundamentally one single layer of graphite; a layer of sp^2 bonded carbon atoms arranged in a honeycomb (hexagonal) lattice. Graphene based nanocomposites with semiconductors have received increasing attention due to their remarkable electrocatalytic, electrochemical sensing and electrochemical energy conversion properties [34].

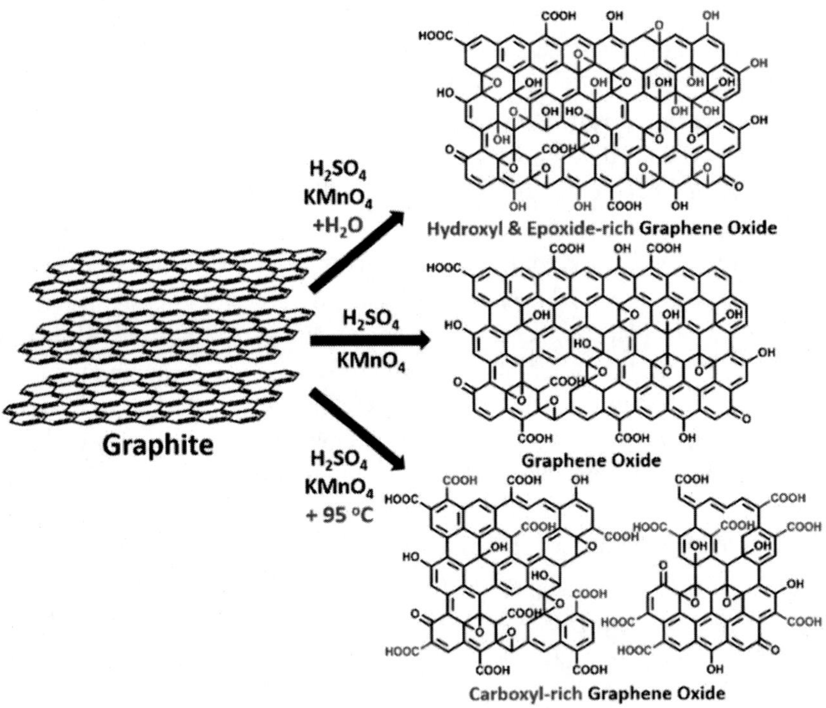

Figure 2. Chemical reduction of Graphite sheets into graphene oxide.

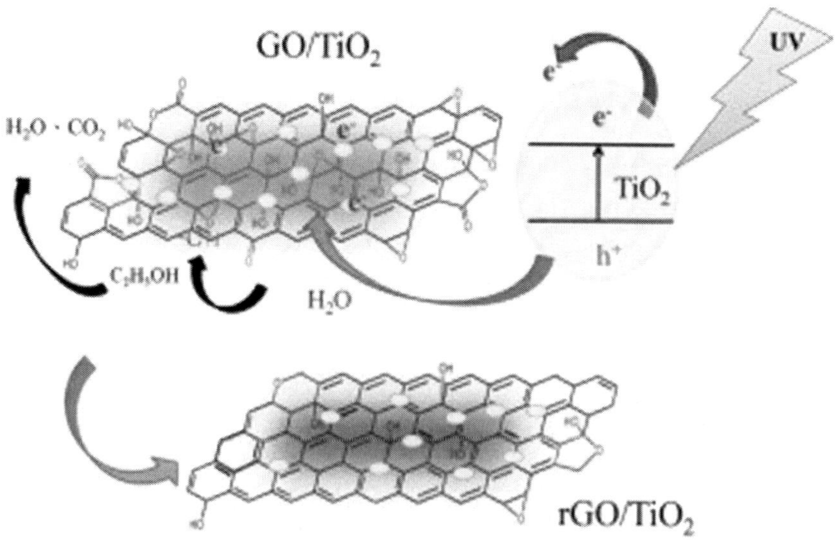

Figure 3. Trapping of photoexcited electron and hole by Graphene oxide and reduced graphene oxide.

The nanocomposite of reduced graphene oxide and titanium dioxide (rGO/TiO_2) is found to have an enhanced photocatalytic effect [35]. Compared to pure titanium dioxide (TiO_2), the enhanced photocatalytic effect of reduced graphene oxide/titanium dioxide nanocomposites can be largely attributed to three factors, which include the following:

1. High quantum efficiency with enhanced electron–hole pair separation.
2. Visible-light harvesting resulting from extended visible-light absorption.
3. High reaction possibility achieved by intensely enhanced reactant absorbability.

The reduced graphene oxide (rGO) and titanium dioxide (TiO_2) heterojunction leads to enhanced separation of the electron–hole pair in titanium dioxide (TiO_2), with the electron being transferred into reduced graphene oxide (rGO) (Figure 3). In this configuration, rGO acts as an electron sink, preventing recombination of excited electrons, while the hole remains in titanium dioxide (TiO_2) to drive the oxidation process [36].

On the other hand, the incorporation of a second metal into titanium dioxide (TiO_2) creates bimetallic photocatalysts with improved photocatalytic activity compared to pure TiO_2. The presence of the second metal can promote charge separation, reduce electron-hole recombination, extend the light absorption range, and enhance the generation and utilization of reactive oxygen species (ROS). This leads to more efficient degradation of pollutants and improved overall photocatalytic performance [37]. Bimetallic titanium dioxide (TiO_2) photocatalysts offer the advantage of tailoring their properties by selecting different metal combinations [38]. Each metal component can introduce unique characteristics, such as band structure modifications, surface properties, and electronic states [39]. These variations enable fine-tuning of the photocatalyst's activity, selectivity, and stability for specific applications. Additionally, the metal ratio in the bimetallic system can be adjusted to optimize the catalytic performance, making it a versatile platform for organic pollutant degradation [40].

The addition of a second metal to titanium dioxide (TiO_2) introduces new functionalities and promotes synergistic effects, resulting in improved photocatalytic activity [41]. Bimetallic titanium dioxide (TiO_2) photocatalysts exhibited higher efficiency in the degradation of various pollutants, including organic dyes, volatile organic compounds (VOCs), and environmental toxins [42]. The presence of the second metal can facilitate the separation of photogenerated electron-hole pairs, reduce electron-hole recombination, and enhance the overall charge transfer process, leading to increased photocatalytic activity [42]. Pure titanium dioxide (TiO_2) primarily absorbs ultraviolet (UV) light, which accounts for only a small fraction of solar radiation. By incorporating a second metal, such as silver (Ag), gold (Au), or platinum (Pt), the bimetallic TiO_2 photocatalyst can extend its light absorption range into the visible spectrum. This broadens the spectrum of light that can be utilized for photocatalytic reactions, making the photocatalyst more efficient under solar irradiation [43].

Bimetallic titanium dioxide (TiO_2) photocatalysts can exhibit selectivity towards specific pollutants, depending on the type of second metal incorporated. Different metals have varying affinities towards specific pollutants, enabling targeted degradation or removal of specific contaminants. This selectivity can be particularly useful in pollutant degradation, the presence of specific pollutants needs to be targeted and eliminated [44].

Hence, this chapter explores the area of alternative environmentally friendly methods for reducing graphene oxide and incorporating the green-synthesized reduced graphene oxide (rGO) into a titania matrix in the presence of other semiconducting metal oxides. This approach can yield excellent reduced graphene oxide (rGO)-based bimetallic heterojunctions, serving as a potent source with pollutant degradation capabilities [45].

2. Study for Alternative Methods for the Reduction of Graphene Oxide Using Nature-Based Reagents

Graphene oxide (GO) is an oxidized form of graphene, containing oxygen-containing groups. Reduced graphene oxide can be obtained from graphene oxide through a reduction process [46]. This can be accomplished through electrochemical reduction, chemical reduction and thermal reduction [47]. Each process has its own merits and demerits but the end goal is same that is to remove the oxygen groups. Chemical reduction of graphene oxide is the most important method for synthesis of reduced graphene oxide [48]. Despite the distinctive advantages, reduction of graphene oxide (GO) via chemical route the reduced graphene oxide (rGO) form irreversible aggregation (resulting from strong van der Waals attractive forces between the graphene planes) which limits its processibility [49]. Another negative aspect of the reduction method is the high toxic nature of the reducing agents (hydrazine, di-methylhydrazine, hydroquinone, sodium borohydride etc.). Thus, to accomplish the objective, the green synthesis principle has prompted researchers worldwide to develop a simple method for green reduction of graphene oxide (GO) as shown in Figure 4; which is readily accessible, sustainable, economical, renewable, and environmentally friendly in nature. Reducing graphene oxide (GO) using nature-based reagents and environment-friendly methods has attracted a lot of scientific interest [50].

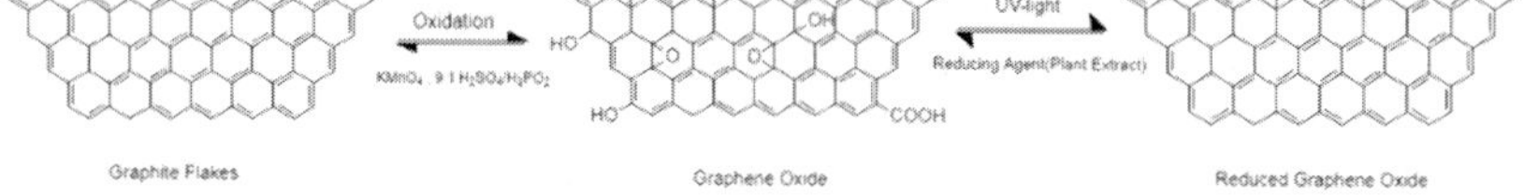

Figure 4. Schematic diagram of green synthesised reduced graphene oxide.

The individual graphene sheets are severely harmed by the conversion of graphite oxide into graphene oxide, which has additional effects upon further reduction of the graphene oxide [51]. Further deterioration of the isolated graphene platelets is not desired because the graphite to graphite oxide oxidation process already reduces their average size [52]. According to the characteristics of the support media, the graphene platelet to platelet contacts can be attenuated by altering the surface functionality in graphene oxide, which contains flakes of monolayer and few layer graphene interleaved with water [53]. Graphite oxide can be transformed into graphene oxide using a variety of techniques [54]. GO is currently primed primarily using Hummers and Offeman's 1958 approach [55]. Here, graphite is reacted with a nonaqueous solution of sodium nitrate and potassium permanganate in concentrated sulfuric acid to produce graphite oxide. While various modifications have been proposed, the fundamental method or strategy remains largely unchanged. Because of this, a brand-new phrase known as the modified Hummers' approach is in use [56].

Hummers and Offeman's most popular methods include sonication, stirring, or a combination of the two. The most efficient method for removing graphite oxide (up to virtually complete exfoliation level) is discovered to be sonication [57]. However, this method also severely damages graphene flakes, causing the surface size to change from microns to nanometers. Another method for producing chemically GO through the Brodie approach [58], the Staudenmaier method [59], the Hummer's method [60], as well as its modifications, such as the modified Hummer's method or the Improved Hummer's method [61], are all conventional techniques. The process used to decrease the parent chemical graphene oxide is very important since it will eventually determine the quality of the reduced graphene oxide (referred to as rGO) that is generated [62]. This in turn will determine how structurally different the rGO is from the original graphene. Large amounts of graphene are required for industrial applications like fuel storage, for which rGO is the most logical answer [63]. As a result, the method used to decrease graphene is crucial [64].

Reduction can be accomplished in a variety of ways using electrochemical, thermogravimetric, or physicochemical methods [65]. In the past, reduced graphene oxide (rGO) was created by heating graphene oxide (GO) in distilled water to various temperatures for varying amounts of time, treating graphene oxide (GO) with hydrazine hydrate utilising hydrogen plasma reduced graphene oxide (rGO), and other methods [66]. Hydrazine hydrate, $NaBH_4$ solution, hydrazine vapour thermal annealing at 900°C and 1100°C with Ar/H_2, and thermal annealing at 1100°C and with vitamin C are the reduction techniques that are typically used to reduce graphene oxide (GO). While some of these techniques can create reduced graphene oxide (rGO) that is of a very high calibre and is comparable to the original graphene, they can also be complicated or time-consuming [67].

The chemical reduction method is a highly accurate and flexible way to reduce graphene oxide (GO) [68]. Unfortunately, research and testing have revealed that the reduced graphene oxide (rGO) that is produced is inferior in terms of surface area and electronic conductivity [69]. When graphene oxide (GO) is reduced at temperatures of 1000°C or higher, reduced graphene oxide (rGO) is created, which has a very high surface area and is extremely similar to the original graphene [70]. Unfortunately, heating causes damage to the graphene platelets, this result in the release of pressure between the carbon dioxide build-up and the gas [71].

Becerril et al. reduced graphene oxide (GO) sheets by thermal annealing at 1000°C and recovered reduced graphene oxide (rGO) using high-quality vacuum at (10^{-5} Torr) [72]. In a commercial microwave oven, graphite oxide particles can also be reduced using microwave and photoreduction, making it simple to obtain reduced graphene oxide (rGO) in just one minute under ambient circumstances [73]. Another intriguing technique for generating reduced graphene oxide (rGO) is the electrochemical elimination of oxygen functions [74]. Unfortunately, all of these procedures only manage to significantly reduce the graphene oxide (GO) mass (by around 30%), which results in serious structural flaws and has a negative impact on the mechanical strength of the reduced graphene oxide (rGO) produced [75]. It has been discovered that photochemical reduction of graphene oxide produces superior-quality reduced graphene oxide with a structure that is extremely close to flawless graphene [76]. Colorants derived from plants or animals can be categorised as natural dyes [77]. The leaves, seeds, fruits, roots, or bark of many plant species contain natural colouring material that is appropriate for use as dye in a variety of businesses [78]. Therefore, using these natural dyes to decrease graphene oxide may also result in the creation of novel, simple graphene chemical approaches [79].

The reduction of graphene oxide can be induced by using natural plant dyes. Visible-light-stimulating dyes are those derived from plants that exhibit high absorption in the visible light spectrum. Rajukumari et al. utilized indigo dye to synthesize reduced graphene oxide by combining graphene oxide (GO) and the dye. A suspension solution was prepared, and it was then used to reduce GO nanosheets [80]. In a typical synthesis, water and indigo plant dye powder are mixed continuously at room temperature to create a dye solution [81]. The dye solution is then combined with a graphene oxide (GO) solution and stirred vigorously for another 30 minutes, resulting in the development of a consistent suspension solution. This suspension solution is then purged with nitrogen gas to remove oxygen, and the (GO + dye) suspension is exposed to visible light for 60 minutes after maintaining the nitrogen gas flow for 30 minutes. As the irradiation time increases, the suspension's color gradually shifts from light yellow to black, indicating the efficient conversion of GO to reduced graphene oxide (rGO). A black precipitate is formed, and the resulting sample is filtered and saved. The required rGO nanopowders are obtained by washing with distilled water and then baking them at 60°C. The visible-light source for the reduction is a 100 W lamp emitting visible light at a wavelength of 400 nm [80, 81].

3. Categorizing Bimetallic Heterojunctions and Mechanism

Whenever two distinct substances with varied work functions, also known as electron affinities, come into electrical contact with one another, a heterojunction is created [82]. The energy levels and band structures of the two materials can vary at such a junction, resulting in intriguing electrical and optical phenomena. The relationship between the semiconductor band edge potentials and the pure metal's Fermi level, as well as the induced electronic states at the interface area, determines the features of the metal-semiconductor interface. The typical photocatalytic heterojunctions can be divided into two categories based on the transfer framework for photogenerated electrons and holes, which is significant. They are type 1 and type 2 photocatalytic heterojunction: Type 1 possesses larger band semiconductors with smaller valence band and conduction band gap levels have their valence and conduction bands enclosed closer. Based on the basic idea of charge carrier transfer, electrons and holes generate energy that will concentrate in smaller band gaps, which will lead to a decline in photocatalytic activity [83].

While for Type 2 heterojunction refers to a band alignment in which the conduction band of a material with a small band gap may be aligned higher than that of a material with a wider gap. The high negative electrons in the semiconductor's larger conduction band level push the electrons towards the semiconductor's smaller conduction band level in this heterojunction. Similar to this, positive holes are transferred to semiconductors with greater energy gaps and less positive valence bands. However type 2 heterojunction possess inevitable leads to reduce redox ability as the electrons and hole that are released exhibit less potent reduction and oxidation potentials than those in the single component photocatalyst, leading to the conclusion that the construction of TiO_2-based type I and II heterojunctions cannot approximately satisfy the requirement of highly efficient photocatalysis [83-85].

The photosynthesis mechanism occurring in plants is quite similar to the Z-scheme photocatalytic process, which involves a two-step photoexcitation pathway as illustrated in Figure 5. In the first step, the electron in the highest occupied molecular orbital (HOMO) of a photosystem II (PSII) stimulated to its lowest unoccupied molecular orbital (LUMO) by solar radiation. Then, the photogenerated electrons in LUMO of PS II are carried to HOMO of ferrodoxin-NAPD reductase, known as photosystem I (PSI) via electron mediator. In addition, electrons in HOMO of PS I are stimulated to LUMO. This gives the excited electrons reside in the LUMO of PS I and the HOMO of PS II, respectively. The excited electrons in the HOMO of PS II are utilized to oxidize H_2O subsequently in the excited electron in the LUMO of PS I are used to reduced carbon dioxide. This scheme of PS I and PSII of photosynthesis process relates to the Z-scheme photocatalytic system depicts potent reduction ability of PS I and the strong oxidation ability of PS II. Therefore, it is expected that the Z-scheme photocatalytic system made from merging two narrow-bandgap semiconductors can attain an excellent redox ability mechanism that is excellent to enhance light absorption [86].

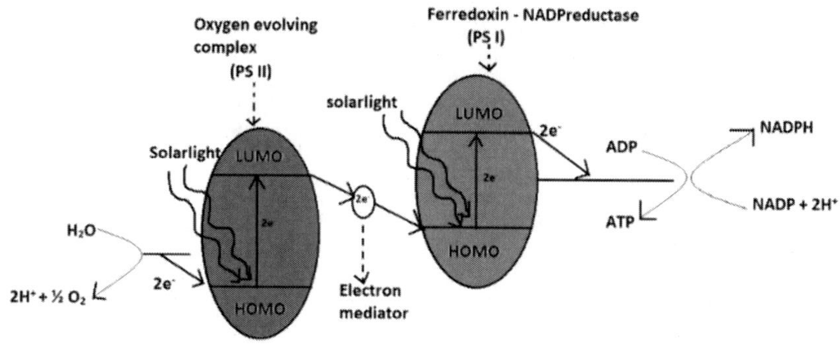

Figure 5. Mechanism of natural photosynthesis using the Z-scheme process.

4. Carbon Components as Electron Mediators in Synthesizing TiO_2 -Based Bimetallic Heterojunctions

TiO_2-based indirect Z-scheme heterojunction systems could also be built using a variety of carbon materials as electron mediators. Jia et al. synthesised a two-step solvothermal synthesis process creating the Z-scheme reduced graphene oxide (rGO), super-thin TiO_2-B nanosheets, and W18O49 nanofibers photocatalyst. TiO_2-B and rGO nanosheets were initially firmly bonded. On rGO/TiO_2-B nanosheets, $W_{18}O_{49}$ nanowires were subsequently produced. The ternary photocatalyst may effectively utilise light for photocatalytic degradation of organic contaminants by absorbing sunlight from the ultraviolet to near infrared areas. With the addition of rGO, both the photon-generated carrier transfer efficiency and light absorption intensity are enhanced. The nonstoichiometric surface plasmon resonance effect gave rise to the localised $W_{18}O_{49}$ is in support of the rGO/TiO_2-B/$W_{18}O_{49}$ composite's absorption range expanding to near infrared light. An electron spin resonance test was used to confirm the Z-scheme rGO/TiO_2-B/$W_{18}O_{49}$ composite's photocatalytic mechanism. Rhodamine B ssolution is degraded photocatalytically using the conduction band of TiO_2-B and the valence band of $W_{18}O_{49}$, which makes it easier to separate photogenerated carriers. Under whole solar spectrum illumination, the rGO/TiO_2-B/$W_{18}O_{49}$ heterostructure displays good photocatalytic performance. The outcome could signify a fresh approach to developing nanocomposites with a suitable band structure for the effective application of full solar light [87].

5. Synthesis of rGO/TiO_2 Bimetallics

The synthesis of rGO/TiO_2 bimetallic heterojunctions involves the integration of reduced graphene oxide (rGO) and titanium dioxide (TiO_2) nanoparticles, with the addition of one or more metallic components. Several methods have been developed to synthesize these hybrid materials, and outlining a commonly used approach as

5.1. Preparation of rGO

Various methods are available for the synthesis of reduced graphene oxide (rGO), including chemical reduction of graphene oxide (GO). One commonly used method involves the reduction of GO using a reducing agent such as hydrazine hydrate or sodium borohydride, followed by thorough washing and purification steps to remove any residual chemicals and by-products. However, in green methods, only a green reducing agent is used, avoiding the use of traditional chemical reducing agents [80].

5.2. Synthesis of TiO_2 Nanoparticles

TiO_2 nanoparticles can be synthesized using various methods, such as sol-gel, hydrothermal, or solvothermal methods. In a typical sol-gel method, titanium precursor compounds (such as titanium alkoxides) are mixed with a solvent and a stabilizing agent. The mixture is then hydrolyzed and thermally treated to form TiO_2 nanoparticles. The resulting TiO_2 nanoparticles can be further characterized and adjusted for their size, shape, and crystallinity through additional post-treatment steps [88].

5.3. Incorporation of Bimetallic Components

The bimetallic components, such as noble metals (e.g., Ag, Au, Pt) or non-noble metals (e.g., Cu, Ni, Fe), can be incorporated onto the surface of TiO_2 nanoparticles through various techniques. One common approach is the deposition-precipitation method, where the TiO_2 nanoparticles are dispersed in a solution containing metal precursor salts. The bimetallic components are then deposited onto the surface of TiO_2 through chemical reduction or other appropriate methods [89].

5.4. Assembly of rGO/TiO_2 Heterojunctions

The final step involves the assembly of the rGO/TiO_2 heterojunctions. This can be achieved by physically mixing the rGO and TiO_2 nanoparticles, followed by sonication to ensure proper dispersion and homogeneity. The mixture is typically subjected to further treatment, such as stirring or refluxing, to enhance the interaction and bonding between rGO and TiO_2 [90].

5.5. Characterization and Optimization

The synthesized rGO/TiO_2 bimetallic heterojunctions should be characterized to determine their structural, morphological, and optical properties. Techniques such as X-ray diffraction (XRD), scanning electron microscopy (SEM), transmission electron microscopy (TEM), UV-Vis spectroscopy, and Fourier-transform infrared spectroscopy (FTIR) can be employed for characterization. The synthesis parameters can be optimized to achieve desired properties, such as the ratio of rGO to TiO_2, the loading of bimetallic components, and the morphology of the heterojunctions [90].

It is important to note that the specific synthesis method may vary depending on the desired application and the bimetallic components being used. The above-described method provides a general framework for the synthesis of rGO/TiO_2 bimetallic heterojunctions in this chapter, which can be modified and tailored according to the specific requirements of the intended application.

6. Applications of rGO/TiO_2 Bimetallics in Organic Pollutant Degradation

6.1 Photocatalytic Degradation of Dyes in Water

One of the pollutants that can now be found in wastewater in the highest concentrations is synthetic dye. This is attributed to the widespread use of dyes in various products, including those used in printing, food and beverage, pharmaceutical, and textile industries [91]. These hazardous and carcinogenic synthetic colors can block sunlight from reaching water bodies, impairing natural aquatic processes, including photosynthesis and other biodegradation procedures. Rhodamine B, methylene blue, and methyl orange are the dyes in photocatalysis that have been investigated the most frequently. These dyes are often used as model compounds to investigate the photocatalytic degradation and transformation process [92]. If organic dyes with complicated chemical structures are released into the environment, they may convert into aromatic amines. These intermediates must be eliminated before entering the environment since they are extremely harmful to aquatic life. However, because dyes are non-biodegradable and persistent, standard wastewater treatment methods are inefficient at degrading them. The photocatalytic process has been suggested as one of the more successful methods. In photocatalysis, the degradation is started by OH radicals attacking the azo link (-N=N-), which is one of the weakest chemical bonds in the dye molecules' chemical structure. The conjugate structure is typically attacked, and it is destroyed [93]. After that, the process's intermediates will eventually undergo a radical chain reaction with the oxygen molecules to generate water and carbon dioxide [94]. The photodegradability of complex-structured dyes is typically low, and the presence of functional groups affects the adsorption properties. Numerous researchers have looked into this while studying the disappearance of different dyes under UV and visible sunlight [95]. Even while photocatalytic degradation has been the subject of numerous studies looking at dye removal, very few of these studies used actual wastewater. Isari et al. (2018) looked at how various anions might co-exist with RhB being removed by photocatalysis utilising the Fe-TiO_2/rGO photocatalyst. The elimination efficiency dropped from 91 to 73%, 78%, 48%, and 57%, respectively, when nitrate, phosphate, chloride, and sulphate anions were present [95]. According to the explanation, chloride and sulphate adsorbed on the photocatalyst more quickly and competed with dye molecules for the ability to scavenge radicals. In contrast, nitrate and phosphate were less adhered to the surface of the photocatalysts. The total organic carbon (TOC) decreased from 930 to 310 mg/L (66.6%) and the COD decreased from 1550 to 634 mg/L (59.1%) after 390 minutes of treating the actual textile wastewater with the same photocatalyst. Additionally, the photocatalytic method was successful in lowering the non-biodegradable organic components in the wastewater as indicated by the rise in carbon oxidation state (COS) and average oxidation state (AOS).

6.2. Photocatalytic Degradation of Organic Pollutant

Another persistent organic contaminant that has received extensive research attention is pesticides. Several major classes of pesticides, insecticides, and herbicides undergo photodegradation. A few studies on the pilot-scale photodegradation of pesticides in wastewater in Spain have been released [96]. According to Zapata et al. (2010), photocatalysis combined with aerobic biological treatment is technically feasible for treating twelve active chemicals found in wastewater on a significant scale. It was advised to use batch photocatalytic biotreatment first, then continuous biotreatment with a reservoir tank in between. The bulk of the active ingredients decomposed fully utilising the combination method, with the exception of pyrimethanil (42.7%) and thiacloprid (84.2%). They observed that the pilot plant's working conditions outside were different from those in the lab and necessitated a longer treatment period and a greater H_2O_2 dosage [97]. Recent interest has centred on visible-light-driven photocatalysis, a new environmental restoration method. This study effectively synthesised a unique $SrCO_3$-$SrTiO_3$/Ag_3PO_4 heterojunction and thoroughly characterised its structure, chemical makeup, and optical performance. $SrTiO_3/Ag_3PO_4$ was also created, but it exhibited certain flaws in its type II heterojunction. The $SrCO_3$-$SrTiO_3/Ag_3PO_4$ active species trapping studies revealed that OH and O_2 participated in the process. $SrCO_3$-$SrTiO_3/Ag_3PO_4$ heterojunction had a photodegradation efficiency that was 1.4 and 2.0 times greater than that of Ag_3PO_4 and $SrCO_3$-$SrTiO_3$, respectively. The photoreaction process of $SrCO_3$-$SrTiO_3/Ag_3PO_4$ might be explained by a brand-new Z-scheme mechanism. The improvement in visible light absorption, the efficient separation of photogenerated charges, and the production of many active species were all factors in the good photocatalytic property [98]. In addition through the use of a unique method called the nanotemplate assisted sol-gel hydrothermal reaction, nanoporous $SrTiO_3$ photocatalysts were created. The SiO_2 nanotemplate in the alkaline-environment hydrothermal reaction helped to control the local reaction environment near the $SrTiO_3$ nanocrystals in addition to acting as a pore generator. As a result, the photocatalyst's surface area and crystallinity were continuously modulated. The degradation of gaseous isopropyl alcohol (IPA) served as a measure of the photocatalytic abilities of the nanoporous $SrTiO_3$ samples. A $SrTiO_3$ sample made by adding 40% template (STO-SiO_2-40%) demonstrated the highest activity because surface area and crystallinity were in their ideal equilibrium. This sample achieves increases in CO_2 evolution of 40 and 8 times when compared to samples made without the template and commercial nano-$SrTiO_3$, respectively. Additionally, a thorough investigation of the IPA photodegradation mechanism over this sample was conducted. Other nanoporous titanates, such as doped $SrTiO_3$ samples and alkali-metal titanates, can also be prepared using this synthetic method [99].

7. Discussion

One of the most common S-S heterojunctions is joining n-type TiO_2 with a p-type semiconductor with a matching energy band structure [100]. Due to its improved separation efficiency of photogenerated e/h^+ pairs with the recombination of inefficient charge carriers, the construction of indirect Z-scheme system between TiO_2 and another semiconductor using noble metals such as Pt, Au, and Ag as electron mediators has attracted increased attention [101-103]. This improves the photocatalytic efficiency, Ag/TiO_2 nanorods and $ZnFe_2O_4$ nanospheres were physically mixed together in a methanol solution while being continuously stirred to create the $ZnFe_2O_4$/Ag/TiO_2

nanocomposite [102]. The stronger connection between 0D $ZnFe_2O_4$ nanospheres and 1D TiO_2 nanorods is advantageous for the transmission of photogenerated electrons and holes at the interface as opposed to the point contact between 0D TiO_2 nanoparticles and 0D $ZnFe_2O_4$ nanospheres [104]. Using a modified solvothermal procedure, Tariq Altalhi and his team have created a gold-titanium (Au@TiO_2) binary nanocomposite anchored in reduced graphene oxide (rGO). The experimental technique was carried out in a single pot step without the use of additional surfactants or capping agents, using only fresh rGO aqueous solution and metal precursors. The solvothermal procedure used DMSO as the solvent. Detailed methods like high-resolution transmission electron microscopy (HRTEM), Raman spectroscopy, and X-ray diffraction powder (XRD) demonstrate the successful production of this heterojunction. Both UV and visible light were interestingly absorbed by the ternary Au@TiO_2/rGO [105]. The ternary nanocomposite demonstrated great efficiency in the photocatalytic degradation of organic pollutant compounds and dyes when compared to pure TiO_2 and binary Au@TiO_2. The broad absorption spectrum of the ternary nanocomposite, which can improve the photogenerated electron-hole pairs and hence, boost the photochemistry reaction at the photocatalyst surface, can be linked to an improvement in photocatalytic performance [105]. Moreover, the rGO layers perform admirably as electron transporters, enabling efficient separation of electron-hole pairs [106]. This chapter focuses on the development of reduced graphene oxide (rGO) materials based on ternary nanostructures for photoactive UV-visible light photocatalysts. Satya Paul and his group reported a series of PdCr bimetallic catalysts with different weight percentage of RGO decorated with TiO_2. It was found that TiO_2 nano catalyst performed exceptionally well for the Suzuki coupling and hydrogenation of nitroarenes and α,β-unsaturated carbonyl compounds as compared to other synthesized catalysts. Introducing Cr^{3+} into the catalyst not only decreased the cost but also enhanced the catalytic activity due to the remarkable synergistic effect between Pd and Cr NPs. These significant synergistic interactions between Pd and Cr account for the observed high catalytic efficacy of the nanocatalyst, which afforded excellent yield in short reaction time and also retain catalytic activity for several runs [107]. Chinnakonda S. Gopinath and his team have reported noble bimetallic nanoparticles (Au–Pd NPs) on a carbon-based support and integrated with Titania to obtain Au–Pd/C/TiO_2 and Au–Pd/rGO/TiO_2 nanocomposites using an eco-friendly hydrothermal method. Here, a 1:1 (w/w) Au–Pd bimetallic composition was dispersed on (a) high-surface-area (3000 m2 g^{-1}) activated carbon (Au–Pd/C), prepared from a locally available plant source (in Assam, India), and (b) reduced graphene oxide (rGO) (Au–Pd/rGO); subsequently, they were integrated with TiO_2 [108]. The shift observed in Raman spectroscopy demonstrated the electronic integration of the bimetal with Titania. The photocatalytic activity of the above materials for the hydrogen evolution reaction was studied under one sun conditions using methanol as a sacrificial agent in a powder form. The photocatalysts were also employed to prepare a thin film by the drop-casting method. Au–Pd/rGO/TiO_2 exhibits 43 times higher hydrogen (H_2) yield in the thin film form (21.50 mmol h^{-1} g^{-1}) compared to the powder form (0.50 mmol h^{-1} g^{-1}). On the other hand, Au–Pd/C/TiO_2 shows 13 times higher hydrogen (H_2) yield in the thin film form (6.42 mmol h^{-1} g^{-1}) compared to the powder form (0.48 mmol h^{-1} g^{-1}). While powder forms of both catalysts show comparable activity, the Au–Pd/rGO/TiO_2 thin film showed 3.4 times higher activity than that of Au–Pd/C/TiO_2. This was ascribed to (a) an effective separation of photogenerated electron–hole pairs at the interface of Au–Pd/rGO/TiO_2 and (b) the better field effect due to plasmon resonance of the bimetal in the thin film form. The catalytic influence of the carbon-based support was highly pronounced due to synergistic binding interaction of bimetallic nanoparticles.

Conclusion

GO can be effectively reduced without the need of any reducing chemicals by using a simple and environmentally safe approach that is based on the naturally occurring dye-stimulated visible-light reduction mechanism. By taking into consideration a viable and robust method for the reduction of GO, this chapter can provide a novel point of view on the development and design of various graphene-based nanomaterials for a range of potential applications. The photo-induced reactions of the Titania semiconducting catalyst are shown to be significantly improved by the addition of GO and rGO to TiO_2.

References

[1] Anandan, S.; Ikuma, Y.; Niwa, K. An overview of semi-conductor photocatalysis: modification of TiO_2 nanomaterials. *Solid state phenomena* 2010, 162, 239-260.

[2] Sun, J.; Xue, H.; Guo, N.; Song, T.; Hao, Y. R.; Sun, J.; Zhang, J. and Wang, Q. Synergetic metal defect and surface chemical reconstruction into $NiCo_2S_4$/ZnS

heterojunction to achieve outstanding oxygen evolution performance. *Angewandte Chemie International Edition* 2021, 60(35), 19435-19441.

[3] Steffi, A. P.; Balaji, R.; Chandrasekar, N.; Prakash, N.; Rajesh, T. P.; Ethiraj, S.; Samuel, M. S. and Vuppala, S. High-performance anti-corrosive coatings based on rGO-SiO_2-TiO_2 ternary heterojunction nanocomposites for superior protection for mild steel specimens. *Diamond and Related Materials* 2022, 125,108968.

[4] Jiang, L.; Zhou, S.; Yang, J.; Wang, H.; Yu, H.; Chen, H.; Zhao, Y.; Yuan, X.; Chu, W.; Li, H. Near-Infrared Light Responsive TiO_2 for Efficient Solar Energy Utilization. *Advanced Functional Materials* 2022, 32(12), 2108977.

[5] Wu, Z.; Yang, P.; Li, Q.; Xiao, W.; Li, Z.; Xu, G.; Liu, F.; Jia, B.; Ma, T.; Feng, S.; Wang, L. Microwave synthesis of Pt clusters on black TiO_2 with abundant oxygen vacancies for efficient acidic electrocatalytic hydrogen evolution. *Angewandte Chemie* 2023, 135(14), 202300406.

[6] Yang, J.; Chen, L.; Xu, J.; Xu, Y.; Wang, W.; Han, B.; Wang, Y. and Zhang, J. Synthesis of cake-like Ti-Bi bimetallic MOFs-derived OV-rich A-TiO_2/β-Bi_2O_3 heterojunctions for photodegradation of ciprofloxacin. *Journal of Alloys and Compounds* 2023, 170277.

[7] Yu, R.; Yang, Y.; Zhou, Z.; Li, X.; Gao, J.; Wang, N.; Li, J. and Liu, Y. Facile synthesis of ternary heterojunction Bi_2O_3/reduced graphene oxide/TiO_2 composite with boosted visible-light photocatalytic activity. *Separation and Purification Technology* 2022, 299, 121712.

[8] Saleh, M. R.; El-Bery, H. M. and Abdelhamid, H. N. Co@ ZIF-8/TiO_2 heterojunction for green hydrogen generation. *Applied Organometallic Chemistry* 2023, 37(3), 6995.

[9] Huang, H. X.; Yuan, J. G.; Zhang, B.; Zhang, J. G.; Zhu, Y. F.; Li, L. Q. and Wu, Y. A noteworthy synergistic catalysis on hydrogen sorption kinetics of MgH_2 with bimetallic oxide Sc_2O_3/TiO_2. *Journal of Alloys and Compounds* 2020, 839, 155387.

[10] Kader, S.; Al-Mamun, M. R.; Suhan, M. B. K.; Shuchi, S. B.; Islam, M. S. Enhanced photodegradation of methyl orange dye under UV irradiation using MoO3 and Ag doped TiO_2 photocatalysts. *Environmental Technology & Innovation* 2022, 27, 102476.

[11] Hossen, M. A.; Solayman, H. M.; Leong, K. H.; Sim, L. C.; Nurashikin, Y.; Abd Aziz, A.; Wu, L.; Monir, M. U. Recent progress in TiO_2-Based photocatalysts for conversion of CO_2 to hydrocarbon fuels: A systematic review. *Results in Engineering* 2022, 100795.

[12] Wang, X.; Li, S.; Chen, P.; Li, F.; Hu, X.; Hua, T. Photocatalytic and antifouling properties of TiO_2-based photocatalytic membranes. *Materials Today Chemistry* 2022, 23,100650.

[13] Robert, D. Photosensitization of TiO_2 by MxOy and MxSy nanoparticles for heterogeneous photocatalysis applications. *Catalysis Today* 2002, 122(1-2), 20-26.

[14] Sampaio, M. J.; Silva, C. G.; Marques, R. R.; Silva, A. M. and Faria, J. L. Carbon nanotube–TiO_2 thin films for photocatalytic applications. *Catalysis Today* 2011, 161(1), 91-96.

[15] Becerril, H. A.; Mao, J.; Liu, Z.; Stoltenberg, R. M.; Bao, Z.; Chen, Y. Evaluation of solution-processed reduced graphene oxide films as transparent conductors. *ACS Nano* 2008, *2*, 463– 470.

[16] O'regan, B. and Grätzel, M. A low-cost, high-efficiency solar cell based on dye-sensitized colloidal TiO_2 films. *Nature* 1991, 353, 737– 740.

[17] Khojasteh, H.; Salavati-Niasari, M.; Safajou, H.; Safardoust-Hojaghan, H. Facile reduction of graphene using urea in solid phase and surface modification by N-doped graphene quantum dots for adsorption of organic dyes. *Diamond Related Material* 2017, 79, 133– 144.

[18] Goswami, P.; Ganguli, J. N. Tuning the band gap of mesoporous Zr-doped TiO_2 for effective degradation of pesticide quinalphos. *Dalton Transactions* 2013, 42, 14480– 14490.

[19] Chen, L.; Tian, L.; Xie, J.; Zhang, C.; Chen, J.; Wang, Y.; Li, Q.; Lv, K.; Deng, K. One-step solid state synthesis of facet-dependent contact TiO_2 hollow nanocubes and reduced graphene oxide hybrids with 3D/2D heterojunctions for enhanced visible photocatalytic activity. *Applied Surface Science* 2020, 144353.

[20] Rajkumari, N. P.; Dolakashoria, S. and Goswami, P. Plant-Based Natural Dye-Stimulated Visible-Light Reduction of GO and Physicochemical Factors Influencing the Production of Oxidizing Species by a Synthesized (rGO)/TiO_2 Nanocomposite for Environmental Remediation. *ACS omega* 2021, 6(4), 2686-2698.

[21] Li, F.; Du, P.; Liu, W.; Li, X.; Ji, H.; Duan, J.; Zhao, D. Hydrothermal synthesis of graphene grafted titania/titanate nanosheets for photocatalytic degradation of 4-chlorophenol: Solar-light-driven photocatalytic activity and computational chemistry analysis. *Chemical Engineering Journal* 2018, 331, 685– 694.

[22] Neelgund, G. M.; Oki, A. Graphene-coupled ZnO: a robust NIR-induced catalyst for rapid photo-oxidation of cyanide. *ACS Omega* 2017, *2*, 9095– 9102.

[23] Smith, A. T.; LaChance, A. M.; Zeng, S.; Liu, B.; Sun, L. Synthesis, properties, and applications of graphene oxide/reduced graphene oxide and their nanocomposites. *Nano Material Science* 2019, 1, 31– 47.

[24] Ipadeola, A. K.; Haruna, A. B.; Abdullah, A. M.; Al-Hajri, R. S.; Viter, R.; Ozoemena, K. I. and Eid, K. Ternary PdNiO nanocrystals-ornamented porous CeO_2/onion-like carbon for electrooxidation of carbon monoxide: unveiling the effect of supports and electrolytes. *Catalysis Science & Technology* 2023.

[25] Saber, N.; Mezni, A.; Alrooqi, A.; Altalhi, T. Fabrication of efficient Au@ TiO_2/rGO heterojunction nanocomposite: Boosted photocatalytic activity under ultraviolet and visible light irradiation. *Journal of Materials Research and Technology* 2021, 2238-2246.

[26] Cheng, G.; Shiyin, Xu.;Peng, X.; Feifan Li.; Yunhui, Xu.; Jin, Y.; Rui, Li.; Hailong, P. Indium Tin-Oxide Wrapped 3D rGO and TiO_2 Composites: Development, Characterization, and Enhancing Photocatalytic Activity for Methylene Blue. *Catalysts* 2019, 9(10), 848.

[27] Viculis, L. M.; Mack, J. J; Mayer, O. M.; Hahn, H. T; Kaner, R. B. Intercalation and Exfoliation Routes to Graphite Nanoplatelets. *Journal of Materials Chem*istry 2005, 15(9), 974.

[28] Razaq, A.; Bibi, F.; Zheng, X.; Papadakis, R.; Jafri, S. H.; Li, H. Review on graphene-, graphene oxide-, reduced graphene oxide-based flexible composites: From fabrication to applications. *Materials* 2022, 15 (3), 1012.

[29] Yeh, T. F.; Cihlář, J.; Chang, C. Y.; Cheng, C. and Teng, H. Roles of graphene oxide in photocatalytic water splitting. *Materials Today* 2013, 16(3),78-84.

[30] Dreyer, D. R.; Park, S.; Bielawski, C. W.; Ruoff, R. S. Reduction of graphite oxide using alcohols. *Chemical Society Reviews* 2010, 39 (1), 228–240.

[31] Suk, J. W.; Piner, R. D.; An, J.; Ruoff, R. S. Mechanical properties of monolayer graphene oxide. *ACS Nano* 2010, 4 (11), 6557–6564.

[32] Olabi, A. G.; Abdelkareem, M. A.; Wilberforce, T.; Sayed, E. T. Application of graphene in energy storage device–A review. *Renewable and Sustainable Energy Reviews* 2021, 135, 110026.

[33] Razaq, A.; Bibi, F.; Zheng, X.; Papadakis, R.; Jafri, S. H.; Li, H. Review on graphene-, graphene oxide-, reduced graphene oxide-based flexible composites: From fabrication to applications. *Materials* 2022, 15 (3), 1012.

[34] Tarcan, R.; Todor-Boer, O.; Petrovai, I.; Leordean, C.; Astilean, S.; Botiz, I. Reduced graphene oxide today. *Journal of Materials Chemistry C* 2020, 8 (4), 1198–122.

[35] Huang, X.; Yin, Z.; Wu, S.; Qi, X.; He, Q.; Zhang, Q.; Yan, Q.; Boey, F. and Zhang, H. Graphene-based materials: synthesis, characterization, properties, and applications. *Small* 2011, 7(14), 1876-1902.

[36] Tu, W.; Zhou, Y. and Zou, Z. Versatile graphene-promoting photocatalytic performance of semiconductors: basic principles, synthesis, solar energy conversion, and environmental applications. *Advanced Functional Materials* 2013, *23*(40), 4996-5008.

[37] Butburee, T.; Bai, Y.; Wang, H.; Chen, H.; Wang, Z.; Liu, G.; Zou, J.; Khemthong, P.; Lu, G. Q. M. and Wang, L., 2018. 2D porous TiO2 single-crystalline nanostructure demonstrating high photo-electrochemical water splitting performance. *Advanced Materials* 2018, 30(21), 1705666.

[38] Wang, Y. Q.; Gu, L.; Guo, Y. G.; Li, H.; He, X. Q.; Tsukimoto, S.; Ikuhara, Y. and Wan, L. J. Rutile-TiO_2 nanocoating for a high-rate $Li_4Ti_5O_{12}$ anode of a lithium-ion battery. *Journal of the American Chemical Society* 2012, 134(18),7874-7879.

[39] Qin, H.; Zhou, G.; Yin, J.; An, L.; Zhao, P.; Wang, T.; Rao, S.; Alireza, K;, Yang, J. and Zhang, L. Enhanced photocatalytic performance of g-C_3N_4@ Ce-Fe bimetallic oxide with Z-scheme heterojunction for rapid degradation of tetracycline and its photodegradation pathway. *Colloids and Surfaces A: Physicochemical and Engineering Aspects* 2022, 652,129780.

[40] Yu, G.; Zhang, Y.; Du, X.; Wu, J.; Liu, C. and Zou, Z. In-situ synthesis of nickel/palladium bimetal/$ZnIn_2S_4$ Schottky heterojunction for efficient photocatalytic hydrogen evolution. *Journal of Colloid and Interface Science* 2022, 623,205-215.

[41] Li, Y.; Li, Z.; Tan, H.; Li, Y.; Tang, F.; Wang, P.; Li, Y.; Liu, H.; Zhao, P. and Liu, Q. Introducing cocatalysts into bimetallic sulfides to construct intercalated dual heterojunctions: A photoelectrochemical sensing platform capable of fast PD-L1 response. *Sensors and Actuators B: Chemical* 2023, 376,133044.

[42] Yoon, J. W.; Kim, J. H.; Jo, Y. M. and Lee, J. H. Heterojunction between bimetallic metal-organic framework and TiO_2: Band-structure engineering for effective photoelectrochemical water splitting. *Nano Research* 2022, 15(9), 8502-8509.

[43] Zhang, Q.; Deng, T. S.; Dou, Y. Q.; Wei, M. Z.; Li, S.; Liu, J. and Cheng, Z. Trimetallic Nanostructures of Silver–Platinum Alloy Shells on Gold Nanorods for Plasmon-Mediated Photocatalysis. *ACS Applied Nano Materials* 2022, 5(11), 17048-17058.

[44] Bathla, A.; Lee, J.; Younis, S. A. and Kim, K. H. Recent advances in photocatalytic reduction of CO_2 by TiO_2–and MOF–based nanocomposites impregnated with metal nanoparticles. *Materials Today Chemistry* 2022, 24, 100870.

[45] Yu, R.; Yang, Y.; Zhou, Z.; Li, X.; Gao, J.; Wang, N.; Li, J. and Liu, Y. Facile synthesis of ternary heterojunction Bi_2O_3/reduced graphene oxide/TiO_2 composite with boosted visible-light photocatalytic activity. *Separation and Purification Technology* 2023, 299,121712.

[46] Geim, A. K.; Novoselov, K. S. The rise of graphene. *Nature Materials* 2007, 6 (3), 183–191.

[47] Kuilla, T.; Bhadra, S.; Yao, D.; Kim, N. H.; Bose, S.; Lee, J, H. Recent Advances in Graphene Based Polymer Composites. *Progress in Polymer Science* 2010, *35* (11), 1350–1375.

[48] Li, J.; Zeng, X.; Ren, T.; Van der Heide, E. The preparation of graphene oxide and its derivatives and their application in bio-tribological systems. *Lubricants* 2014, 2(3), 137–161.

[49] Dreyer, D. R.; Todd, A. D.; Bielawski, C. W. Harnessing the Chemistry of Graphene Oxide. *Chemical Society Reviews* 2014, 43(15), 5288.

[50] Weng, X.; Wu, J.; Ma, L.; Owens, G.; Chen, Z. Impact of Synthesis Conditions on Pb(II) Removal Efficiency from Aqueous Solution by Green Tea Extract Reduced Graphene Oxide. Chemical Engineering Journal 2019, 359, 976–981.

[51] Yu, Z. R.; Mao, M.; Li, S. N.; Xia, Q. Q.; Cao, C. F.; Zhao, L.; Zhang, G. D.; Zheng, Z. J.; Gao, J. F. and Tang, L. C. Facile and green synthesis of mechanically flexible and flame-retardant clay/graphene oxide nanoribbon interconnected networks for fire safety and prevention. *Chemical Engineering Journal* 2021, 405, 126620.

[52] Ray, S. C. Application and Uses of Graphene Oxide and Reduced Graphene Oxide. *Applications of Graphene and Graphene-Oxide Based Nanomaterials* 2015, 39–55.

[53] Fan, H.; Wang, L.; Zhao, K.; Li, N.; Shi, Z.; Ge, Z.; Jin, Z. Fabrication, Mechanical Properties, and Biocompatibility of Graphene-Reinforced Chitosan Composites. *Biomacromolecules* 2010, 11 (9), 2345–2351.

[54] Romero, A.; Lavin-Lopez, M. P.; Sanchez-Silva, L.; Valverde, J. L.; Paton-Carrero, A. Comparative Study of Different Scalable Routes to Synthesize Graphene Oxide and Reduced Graphene Oxide. *Materials Chemistry Physics* 2018, 203, 284–292

[55] Stankovich, S.; Dikin, D. A.; Piner, R. D.; Kohlhaas, K. A.; Kleinhammes, A.; Jia, Y.; Wu, Y.; Nguyen, S. T.; Ruoff, R. S. Synthesis of Graphene-Based Nanosheets via Chemical Reduction of Exfoliated Graphite Oxide. *Carbon* 2007, 45 (7), 1558–1565

[56] Agarwal, V.; Zetterlund, P. B. Strategies for Reduction of Graphene Oxide – a Comprehensive Review. *Chemical Engineering Journal* 2021, 405, 127018.

[57] Hummers Jr, W. S. and Offeman, R. E. Preparation of graphitic oxide. *Journal of the american chemical society* 1958, 80(6), 1339-1339.

[58] Darabdhara, G. and Das, M. R. Bimetallic Au-Pd nanoparticles on 2D supported graphitic carbon nitride and reduced graphene oxide sheets: A comparative photocatalytic degradation study of organic pollutants in water. *Chemosphere* 2018, 197, 817-829.

[59] Brodie, B. C. On the atomic weight of graphite. *Philosophical transactions of the Royal Society of London* 1859, (149),249-259.

[60] Staudenmaier, L. Verfahren zur darstellung der graphitsäure. *Berichte der deutschen chemischen Gesellschaft* 1898, 31(2),1481-1487.

[61] Hummers Jr, W. S. and Offeman, R. E. Preparation of graphitic oxide. *Journal of the american chemical society* 1958, 80(6),1339-1339.

[62] Marcano, D. C.; Kosynkin, D. V.; Berlin, J. M.; Sinitskii, A.; Sun, Z.; Slesarev, A.; Alemany, L. B.; Lu, W.; Tour, J. M. Improved synthesis of graphene oxide. *ACS nano* 2010, 4(8), 4806-4814.

[63] Singh, P. K.; Singh, U.; Bhattacharya, B.; Rhee, H. W. Electrochemical Synthesis of Graphene Oxide and Its Application as Counter Electrode in Dye Sensitized Solar Cell. *Journal of Renewable and Sustainable Energy* 2014, 6 (1), 013125.

[64] Ossonon, B. D.; Bélanger, D. Synthesis and Characterization of Sulfophenyl-Functionalized Reduced Graphene Oxide Sheets. *RSC Advances* 2017, *7* (44), 27224–27234.

[65] Yu, W.; Sisi, L.; Haiyan, Y.; Jie, L. Progress in the functional modification of graphene/graphene oxide: A review. *RSC advances* 2020, 10(26)15328-15345.

[66] Biswas, H. S., Bala, S., Kundu, A. K., Saha, I., Poddar, S., Sarkar, S. and Mandal, P. Tuned synthesis and designed characterization of graphene oxide thin film. *Inorganic Chemistry Communications* 2022, 139,109356.

[67] Li, Z.; Song, E.; Ren, R.; Zhao, W.; Li, T.; Liu, M.; Wu, Y. Pd–Pd/PdO as active sites on intercalated graphene oxide modified by diaminobenzene: fabrication, catalysis properties, synergistic effects, and catalytic mechanism. *RSC advances* 2022, 12(14), 8600-8610.

[68] Xu, K.; Wang, L.; Zhang, Y.; Tang, H.; Zhu, L.; Zhao, D.; Yan, Z.; Cao, X. Graphene oxide enabled self-assembly of silver trimolybdate nanowires into robust membranes for nanosolid capture and molecular separation. *Nanoscale 2023*, 15(14), 6607-6618.

[69] Immanuel, S. and Sivasubramanian, R. Fabrication of two-dimensional chemically reduced graphene oxide nanosheets for the electrochemical determination of epinephrine. *Bulletin of Materials Science* 2020, *43*(1),79.

[70] Habte, A. T. and Ayele, D. W. Synthesis and characterization of reduced graphene oxide (rGO) started from graphene oxide (GO) using the tour method with different parameters. *Advances in Materials Science and Engineering*, 2019.

[71] Li, Y.; Zhao, W.; Weyland, M.; Yuan, S.; Xia, Y.; Liu, H.; Jian, M.; Yang, J.;Easton, C. D.; Selomulya, C. and Zhang, X., 2019. Thermally reduced nanoporous graphene oxide membrane for desalination. *Environmental science & technology* 2019, 53(14), 8314-8323.

[72] Ngqalakwezi, A.; Nkazi, D.; Seifert, G. and Ntho, T. Effects of reduction of graphene oxide on the hydrogen storage capacities of metal graphene nanocomposite. *Catalysis Today* 2020, 358, 338-344.

[73] Pei, S. and Cheng, H. M. The reduction of graphene oxide. *Carbon* 2012, 50(9), 3210-3228.

[74] Sharma, N.; Arif, M.; Monga, S.; Shkir, M.; Mishra, Y. K. and Singh, A. Investigation of bandgap alteration in graphene oxide with different reduction routes. *Applied Surface Science* 2020, 513, 145396.

[75] Zuo, K.; Huang, X.; Liu, X.; Gil Garcia, E. M.; Kim, J.; Jain, A.; Chen, L.; Liang, P.; Zepeda, A.; Verduzco, R. and Lou, J. A hybrid metal–organic framework–reduced graphene oxide nanomaterial for selective removal of chromate from water in an electrochemical process. *Environmental Science & Technology* 2020, 54(20), 13322-13332.

[76] Pal, P.; Yadav, A.; Chauhan, P. S.; Parida, P. K. and Gupta, A. Reduced graphene oxide based hybrid functionalized films for hydrogen detection: Theoretical and experimental studies. *Sensors International* 2021, 2, 100072.

[77] Alvi, I. H.; Li, Q.; Hou, Y.; Onyekwena, C. C.; Zhang, M. and Ghaffar, A. A critical review of cement composites containing recycled aggregates with graphene oxide nanomaterials. *Journal of Building Engineering* 2023,105989.

[78] Jin, X.; Li, N.; Weng, X.; Li, C.; Chen, Z. Green Reduction of Graphene Oxide Using Eucalyptus Leaf Extract and Its Application to Remove Dye. *Chemosphere* 2018, 208, 417–424.

[79] Li, C.; Zhuang, Z.; Jin, X.; Chen, Z. A Facile and Green Preparation of Reduced Graphene Oxide Using Eucalyptus Leaf Extract. *Applied Surface Science* 2017, 422, 469–474.

[80] Akhavan, O.; Kalaee, M.; Alavi, Z. S.; Ghiasi, S. M.; Esfandiar, A. Increasing the Antioxidant Activity of Green Tea Polyphenols in the Presence of Iron for the Reduction of Graphene Oxide. *Carbon* 2012, 50 (8), 3015–3025.

[81] Rajkumari, N. P.; Rouf, A.; Dutta, P.; Goswami, P. Synchronizing charge-carrier capacity and interfacial morphology of green rGO modified ZnO and TiO_2 heterojunctions and study of their photocatalytic behaviour towards UV and visible light active drug and dye. *Materials Science and Engineering*: B 2023, 287, 116094.

[82] Wang, Y.; Shi, Z. X.; Yin. J. Facile Synthesis of Soluble Graphene via a Green Reduction of Graphene Oxide in Tea Solution and Its Biocomposites. *ACS Appl Mater Interfaces* 2011, 3 (4), 1127–1133.

[83] Mills, A. and Le Hunte, S., 1997. An overview of semiconductor photocatalysis. *Journal of photochemistry and photobiology A: Chemistry* 1997, 108(1),1-35.

[84] Pirhashemi, M.; Habibi-Yangjeh, A.; Pouran, S. R. Review on the criteria anticipated for the fabrication of highly efficient ZnO-based visible-light-driven photocatalysts. *Journal of industrial and engineering chemistry* 2018, 62, 1-25.

[85] Zhang, L.; Jaroniec, M. Toward designing semiconductor-semiconductor heterojunctions for photocatalytic applications. *Applied Surface Science* 2018, 430, 2-17.

[86] Maeda, K. Z-scheme water splitting using two different semiconductor photocatalysts. *ACS Catalysis* 2013, 3, 1486–1503.

[87] Zhou, P.; Yu, J.; Jaroniec, M. All-solid-state Z-scheme photocatalytic systems. *Advanced Materials* 2014, 26, 4920-4935.

[88] Jia, C.; Zhang, X.; Matras-Postolek, K.; Huang, B.; Yang, P. Z-scheme reduced graphene oxide/TiO_2-Bronze/$W_{18}O_{49}$ ternary heterostructure towards efficient full solar-spectrum photocatalysis. *Carbon* 2018, 139, 415-426.

[89] Sarika, S.; Abhilash, S.; Sumi, V. S. and Rijith, S. Graphene oxide supported transition metal mixed oxide nanourchins onto bimetallic phosphide coatings as high performance hydrogen evolution electrodes in alkaline media. *Journal of Alloys and Compounds* 2021, 875, 160033.

[90] Li, Z.; Yang, D.; Chu, H.; Guo, L.; Chen, T.; Mu, Y.; He, X.; Zhong, X.; Huang, B.; Zhang, S. and Gao, Y. Efficient Charge Transfer Channels in Reduced Graphene Oxide/Mesoporous TiO_2 Nanotube Heterojunction Assemblies toward Optimized Photocatalytic Hydrogen Evolution. *Nanomaterials* 2022, 12(9), 1474.

[91] Šíma J.; Hasal P. Photocatalytic degradation of textile dyes in a TiO_2 /UV system. *Chemical Engineering Transactions* 2013, 32, 79–84.

[92] Hickman, R.; Walker, E.; Chowdhury, S. TiO_2-PDMS composite sponge for adsorption and solar mediated photodegradation of dye pollutants. *Journal of water process engineering* 2018, 24, 74-82.

[93] Ye, Y.; Feng, Y.; Bruning, H.; Yntema, D.; Rijnaarts, H. H. M. Photocatalytic degradation of metoprolol by TiO_2 nanotube arrays and UV-LED: Effects of catalyst properties, operational parameters, commonly present water constituents, and photo-induced reactive species. *Applied Catalysis B: Environmental* 2018, 220, 171-181.

[94] Saravanan, R.; Gracia, F.; Stephen, A. Basic principles, mechanism, and challenges of photocatalysis. *Nanocomposites for visible light-induced photocatalysis* 2017, 19-40.

[95] Isari, A. A.; Payan, A.; Fattahi, M.; Jorfi, S.; Kakavandi, B. Photocatalytic degradation of rhodamine B and real textile wastewater using Fe-doped TiO_2 anchored on reduced graphene oxide (Fe-TiO_2/rGO): Characterization and feasibility, mechanism and pathway studies. *Applied Surface Science* 2018, 462, 549-564.

[96] Zapata, A.; Oller, I.; Sirtori, C.; Rodríguez, A.; Sánchez-Pérez, J. A.; López, A.; Mezcua, M.; Malato, S. Decontamination of industrial wastewater containing pesticides by combining large-scale homogeneous solar photocatalysis and biological treatment. *Chemical Engineering Journal* 2010, 160, 447-456.

[97] Chen, Y.; Peng, Z.; Li, Y.; Liu, Y.; Chen, Y.; Wu, Y.; Xu, R.; Wang, S.; Zeng, Z. Photocatalytic performance of Z-scheme SrCO 3-SrTiO $_3$/Ag $_3$ PO $_4$ heterojunction for tetracycline hydrochloride degradation. *Journal of Materials Science* 2021, 56, 4356-4365.

[98] Ouyang, S.; Li, P.; Xu, H.; Tong, H.; Liu, L.; Ye, J. Bifunctional-nanotemplate assisted synthesis of nanoporous $SrTiO_3$ photocatalysts toward efficient degradation of organic pollutant. *ACS Applied Materials & Interfaces* 2014, 6, 22726-22732.

[99] Kondo, K.; Murakami, N.; Ye, C.; Tsubota, T.; Ohno, T. Development of highly efficient sulfur-doped TiO_2 photocatalysts hybridized with graphitic carbon nitride. *Applied Catalysis B: Environmental* 2013, 142, 362-367.

[100] Cui, X.; Li, H.; Yang, Z.; Li, Y.; Zhang, P.; Zheng, Z.; Wang, Y.; Li, J.; Zhang, X. A novel $CaIn_2S_4/TiO_2$ NTAs heterojunction photoanode for highly efficient photocathodic protection performance of 316 SS under visible light. *Nanotechnology* 2021, 32, 395702.

[101] Wang, C.; Zhang, W.; Fan, J.; Sun, W.; Liu, E. S-scheme bimetallic sulfide $ZnCo_2S_4$/g-C_3N_4 heterojunction for photocatalytic H_2 evolution. *Ceramics International* 2021, 47, 30194-30202.

[102] Tahir, M. Well-designed $ZnFe_2O_4/Ag/TiO_2$ nanorods heterojunction with Ag as electron mediator for photocatalytic CO_2 reduction to fuels under UV/visible light. *Journal of CO_2 Utilization* 2020, 37, 134-146.

[103] Subha, N.; Mahalakshmi, M.; Myilsamy, M.; Neppolian, B.; Murugesan, V. Direct Z-scheme heterojunction nanocomposite for the enhanced solar H_2 production. *Applied Catalysis A: General* 2018, 553, 43–51.

[104] Saber, N. B.; Mezni, A.; Alrooqi, A.; Altalhi, T. Fabrication of efficient Au@ TiO_2/rGO heterojunction nanocomposite: Boosted photocatalytic activity under ultraviolet and visible light irradiation. *Journal of Materials Research and Technology* 2021, 12, 2238-2246.

[105] Kondo, K.; Murakami, N.; Ye, C.; Tsubota, T.; Ohno, T. Development of highly efficient sulfur-doped TiO_2 photocatalysts hybridized with graphitic carbon nitride. *Applied Catalysis B: Environmental* 2013, 142, 362-367.

[106] Nasr, M.; Balme, S.; Eid, C.; Habchi, R.; Miele, P.; Bechelany, M. Enhanced visible-light photocatalytic performance of electrospun rGO/TiO_2 composite nanofibers. *The Journal of Physical Chemistry C* 2017, 121, 261-269.

[107] Sharma, N.; Sharma, C.; Sharma, S.; Sharma, S.; Paul, S. The synergetic effect of PdCr based bimetallic catalysts supported on RGO-TiO_2 for organic transformations. *Results in Chemistry* 2022, 4, 100524.

[108] Tudu, B.; Nalajala, N.; Reddy, K.; Saikia, P.; Gopinath, C. Electronic integration and thin film aspects of Au–Pd/rGO/TiO_2 for improved solar hydrogen generation. *ACS Appl. Mater. Interfaces* 2019, 11, 36, 32869–32878.

Chapter 8

Synthesis and Applications of Bimetals and Bimetal Oxides for Wastewater Treatment

Paran Jyoti Borah
Sanjyokta Borah
Tejaswini Kalita
Uddipta Kolita
and Rekha Rani Dutta*
Department of Chemistry, The Assam Kaziranga University, Koraikhowa, Jorhat, India

Abstract

Water pollution deteriorates the condition of water shortage by contaminating large volumes of existing water resources. There has been a growing interest in the usage of bimetallic nano particles and bimetallic oxides for the elimination of pollutant in waste water in recent development. This chapter aims to provide a comprehensive overview of the synthetic methodologies for the bimetals and bimetal oxides as documented in current literature. It also summarizes the miscellaneous applications of bimetal and bimetal oxide nanoparticles in the waste water treatment focusing primarily on the removal of heavy metals, chlorinated pesticides, nitro compounds, azo dyes and other anions and oxyanions.

Keywords: bimetals, bimetal oxides, synthesis, application, waste water

---

* Corresponding Author's Email: rekha@kazirangauniversity.in.

In: Bimetals
Editors: Pranjal Saikia, Jayanta Kumar Sarmah and Parikshit GogoiName
ISBN: 979-8-89113-496-6

1. Introduction

The necessity for adequate water, food and energy has itself become a challenge to mankind. It is approximated that by the year, 2030, there will be a need of 30% more water, 40% more food and 50% more energy. The linear production methods implemented to achieve large scale resource goals lead to significant pollution issues related to soil, water and air. These problems stem from the unsustainable extraction of natural resources, as well as inefficient usage and disposal practices. Since all these factors are closely related to our environment, they can directly contribute to environmental pollution and pose health hazards. As a consequence, the development of clean water technologies has become one of the key important agenda of the sustainability policies adopted by various national governments and international agencies. For example, the 6th goal of India's recently prepared sustainable development policy focuses on ensuring access to clean water and sanitation. In recent times, the incorporation of toxic pollutants through food and water and absorption through skin has become a serious threat to public health.

Among the various water purification technologies, water treatment technologies stand out as efficient solutions designed to accelerate the natural processes of water purification. It is evident that with increasing amount of wastewater and high population density, the natural purification process may become overloaded. It is reported that globally, 80% of the wastewater is discharged without treatment. To safeguard both the planet and human health, it is extremely important to purify water as much as possible inorder to maintain a green environment.

In recent years, bimetals and bimetal oxides based technologies have emerged as potential candidates in waste water treatment. Bimetals and bimetal oxides are materials made from two or more different metals or metal oxides, respectively. They can be synthesized through various physical and chemical methods. The reason behind the wider applications of bimetallic systems in areas like catalysis, electrochemistry, metallurgy and microelectronics can be attributed to their characteristic bimetallic surface bonding and the extensive variation in metal electron density. Bimetallic materials are more efficient than their parallel precursors owing to their excellent synergic effect, unique physical and chemical properties and different modes of action [1].

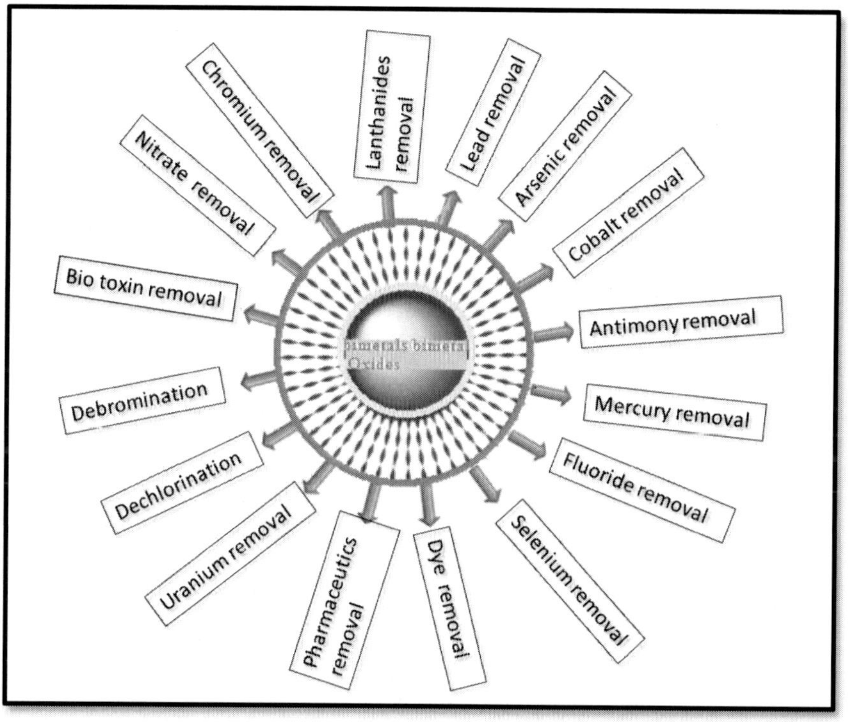

Figure 1. Schematic diagram indicating the use of bimetals and bimetal oxides in water purification.

Bimetallic oxides based catalysts are used in wastewater treatment to facilitate the oxidation as well as reduction of pollutants. These catalysts typically consist of two metal ions, one is the host and the other one plays the functional role. The bimetallic nature of these catalysts increases their activity and stability, which makes them more effective at breaking down organic pollutants and heavy metals. Bimetallic oxide catalysts, often used in advanced oxidation processes like the Fenton reaction which utilizes hydrogen peroxide to oxidize pollutants, are a promising technology for enhancing wastewater treatment efficiency and effectiveness. There are a variety of waste water treatment technologies that are currently in use or in development. These include physical, chemical, and biological processes. Additionally, the properties of bimetals and bimetal oxides are often different than those of the individual metals or metal oxides, which can make them useful for specific applications. For example, bimetals can exhibit improved thermal stability, catalytic activity, and mechanical properties, while bimetal oxides can exhibit improved electronic and optical properties.

Physical processes such as sedimentation, filtration, and flotation are used to remove solid particles from waste water. Chemical processes such as coagulation, flocculation, and chlorination are used to remove specific pollutants. Biological processes such as activated sludge, aeration, and bio filtration use microorganisms to break down organic matter and remove nutrients. Advanced treatment technologies such as membrane filtration, reverse osmosis, and ultraviolet disinfection are also being used to remove a wider range of pollutants. Another recent trend is the development of decentralized treatment systems that can be installed on-site at individual homes, businesses, or small communities, rather than relying on large centralized treatment plants. This allows for greater flexibility and can be more cost-effective for certain areas. Bimetallic and bimetallic oxide catalysts, be effective in treating waste water by catalyzing the oxidation, reduction of pollutants like metals, heavy metals, however, there are several challenges associated with their use in waste water treatment in the context of their stability, selectivity, toxicity, complexity, cost and recovery. Overall, bimetallic and bimetallic oxide catalysts are promising candidates for waste water treatment, but further research is needed to address these challenges and make them more efficient for widespread use. The overarching goal of wastewater treatment is to remove pollutants for safe water reuse or environmental discharge. With the growing challenges of increasing population, urbanization, and climate change, the demand for advanced and efficient wastewater treatment technologies is ever more critical [2].

This chapter provides an overview of the different synthetic routes and methods for the bimetals and bimetal oxides. A brief summary of the different synthesis methods of bimetal and bimetal oxides available in literature are presented with a future trend as well as, their diverse applications in waste water treatment are discussed.

2. Synthesis of Bimetallic Nanoparticles

In order to synthesize bimetallic nanoparticles with the desired shape and characteristics that are adequate for the waste water treatment, it is crucial to select the appropriate approach and experimental processes [3]. The mode of synthesis also affects the catalytic activity of the particles. Figure 2 summarizes the different routes of synthesis of bimetallic nanoparticles.

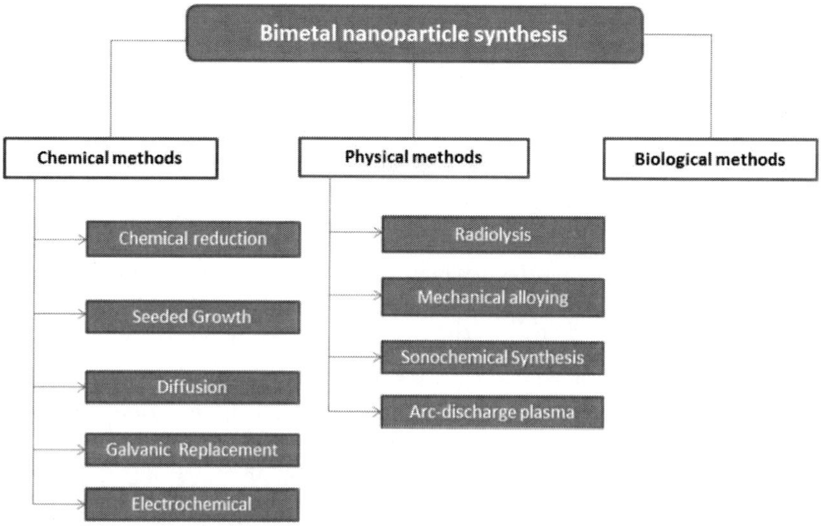

Figure 2. Different approach of synthesis of bimetallic nanoparticles.

2.1. Chemical Methods

2.1.1. Chemical Reduction

In this method, metal salts are reduced chemically with a suitable reductant using an appropriate solvent. It is the most common technique for the synthesis of bimetallic nanoparticles [4]. Here two different types of metals are reduced in a solution employing coreduction method to control the structure of bimetals. The metal ion having highest redox potential reduces first and forms the core followed by the reduction and deposition of the second metal to form the core-shell nanoparticles [5]. However, if the two component metal ions have a high enthalpy of mixing and a comparable redox potential, then simultaneous reduction will occur resulting in the formation of alloyed bimetallic particles [5]. By adjusting the experimental conditions, such as the strength of the reducing agent, the reduction potentials of the metal ions, the reaction temperature, the type of coordinating ligands, and the capping agent, the co-reduction strategy can be used to produce bimetallic particles with a variety of structures [6]. Nevertheless, using strong reductants can make the control of reduction and nucleation processes challenging due to the differences in reduction potential and chemical behavior of the two metals.

This issue can be overcome by utilizing appropriate surfactants and weaker reducing agents [6].

2.1.2. Seed-Mediated Growth

A notable method for creating core-shell and heterostructure bimetallic compounds is the seeded growth technique. The core-shell structures l form when the secondary metal deposits on the surface of the seed metal, whereas the heterostructures are formed when the secondary metal deposits on some specific sites of the seed [6]. As a result, in the seeded growth pathway, the growth mode of the second metal on the seed metal determines the structure of the bimetallic nanocrystals [7]. It is one among the former synthetic routes for generating bimetallic nanoparticles. By utilizing this synthetic technique, a wide range of morphological variants, such as core-shells, alloys, dendritic structures, and asymmetric bimetallic nanocrystals can be synthesized [6]. The synthesis of Ag-Zn bimetallic system was reported by Dabwan and his co-workers in 2010. The silver ions undergo fast reduction to metallic silver by the oxidation of Zn metal and get precipitated on the zinc surface to produce the Ag-Zn bimetallic particles [8]. Bimetallic systems like Pd–Fe, Ni–Fe, Ni–Zn, Cu–Zn, Ag-Fe, Cu-Al, Cu-Fe were also synthesized following this method [9].

2.1.3. Diffusion Method

To create bimetals with a homogenous structural morphology like the core-shell bimetallic nanoparticles, diffusion method is often considered as an excellent choice. Compared with wet chemical reduction methods, diffusion can achieve the aim more easily and mildly. According to the report, precursors H_2PdCl_4 and $HAuCl_4$ diffused to create trisoctahedral Au/Pd alloy nanocrystals. The work discovered that, after the reaction temperature reached a particular level, the rate of precursor diffusion to the surfaces of the nuclei was the single factor affecting how quickly each component's crystals grew [9].

2.1.4. Galvanic Replacement

Oxidation of one metal ion by other metal ion having more positive reduction potential then the former one result in the generation of a galvanic couple. Here initial seeds serve as a reductant for the second metal precursor in the galvanic replacement reaction pathway. Some seeded metal atoms on the surface diffuse into solution by losing electrons if the redox potential of the seeded metal is lower than that of a second metal. These electrons are taken up by the second metal precursors, which then reduce and deposit on the seed metal's surface. The locations of the dissolution and deposit are closely connected to the surface capping agents used [9].

2.2. Physical Methods

In this approach, the interaction between the metallic components is accomplished by physical means typically through radiolysis, mechanical alloying, sonochemical and arc discharge plasma method.

2.2.1. Radiolysis

Here an aqueous solution containing stabilizer ligands and metal precursors is subjected to gamma radiations, which generates solvated electrons. These solvated electrons which are produced due to the dissociation of water molecules carry out the process of reduction of metallic ions into bimetallic nanoparticles. The size and structure of the so-prepared nanoparticles are associated with the dosage of metal precursors. The rate and final nano-structure that the bimetallic particles formed are dependent on the γ-ray dosage. With higher doses, homogenous alloys with core-shell morphologies are more likely to develop throughout this synthesis method. [10] used a straightforward, environmentally friendly (green) process to create bimetallic Pt-Au and Pt-Fe nanoparticles in aqueous solutions without the use of any chemicals other than the respective metal salts and the dispersing agent. In this synthesis he only used a high intensity laser. So it can be concluded that the physical processes are more environmental friendly compared to chemical method.

2.2.2. Mechanical Alloying Method

A convenient and effective method of treating solid state powders is mechanical alloying, which involves repeated cold welding and breaking powder particles in a high-energy ball mill. The bimetallic nanoparticles synthesized by this method generally used in the processes like hydrogen generation, de-chlorination and hydrogenation etc. [11]. Due to the wide range of applications and the sustained solution-free functioning, there is currently active research going on in this area of synthesis [12]. Due to its quick operation time and simplicity, bimetallic systems like Fe-Al, Co-Zn etc. are synthesized using this method [13, 14].

2.2.3. Sonochemical Synthesis

In sonochemical synthesis metal ions are reduced by using high intensity ultrasound. It is a very valuable method for the synthesis of bimetallic nanoparticles. Mizukoshi et al. 2000 created Au_{core}-Pd_{shell} bimetallic nanoparticles by applying high-intensity ultrasound to an aqueous solution of gold (III) and palladium (II) ions. They employed sodium dodecyl sulphate as a reducing agent and a stabiliser for the nanoparticles. The progressive reduction of gold followed by Pd was achieved by ultrasonication carried out in a temperature-controlled bath (approximately 20ºC) with the ultrasonicator operating at 200 kHz and input power of 200 W [12].

2.2.4. Arc-Discharge Plasma

It is an electrical method of synthesis. This method involves the generation of plasma by the arc discharged from direct or alternate or pulsed current. Ag-Cu and Ni-Fe bimetallic nanoparticles are synthesized by utilizing this method [12].

2.3. Biological Methods

Besides, above mentioned procedures, a third method of synthesis of bimetallic nanoparticles is reported in some literatures. This biosynthetic approach for the manufacture of bimetallic nanoparticles employs secure and environmentally acceptable reducing agents, such as plant extracts. It is superior to the chemical and physical methods in many aspects. The physical techniques often have lower production rates and large maintenance costs for maintaining intense reaction conditions [15, 16], while using different chemicals to accomplish reduction, high yield, low-cost chemical techniques face limitations from hazardous solvents, the toxicity of the precursor chemicals, and the production of more toxic by-products [17]. So a green procedure like biological method for preparation of bimetals definitely plays a vital role. The method comprises two routes. The biological reduction of the metal ions into stable forms enables the first process. The second mechanism involves the conversion of metal ions into their stable forms after they have bonded with organisms.

3. Synthesis of Bimetallic Oxide

Spray pyrolysis, impregnation, hydrothermal, precipitation, and sol-gel procedures are common techniques for creating bimetallic oxides (Figure 3). The performance of the synthesized bimetallic oxides is often influenced by variables including calcination temperature, depositional path, the choice and quantity of chemical precursors, and the composition of the solvent.

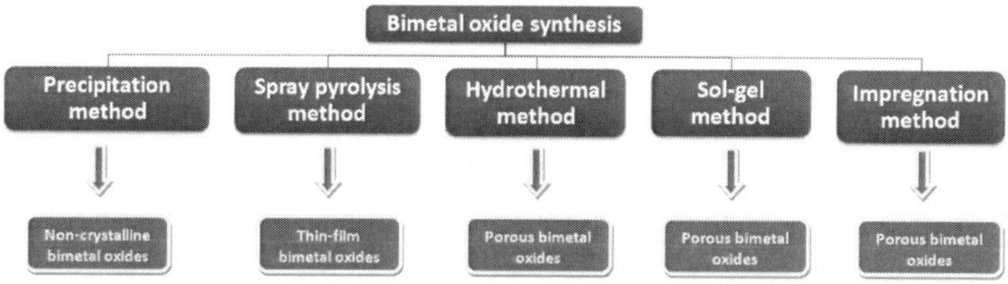

Figure 3. Different approaches for bimetal oxide synthesis.

3.1. Precipitation Method

A wet chemical process called precipitation is used to create non-crystalline nano bimetal oxides. Sodium hydroxide and ammonia are employed as precipitating agents [18, 19]. The nanoparticle size is controlled by maintaining the calcination process at the ideal temperature and then carefully dividing the sediment. The agglomeration is eliminated by the addition of stabilizing agents and dispersants. Precipitation was used to create the bimetal oxides of Fc-Mn, Ni-Mn, and Fe-Ti [20, 21] are some reported systems which can be utilized in the wastewater treatment.

3.2. Spray Pyrolysis Method

An effective, repeatable, and affordable technique for creating bimetal oxide thin films and nano powders from their metal precursor gel solutions is spray pyrolysis. Bimetallic precursors were sprayed and deposited on substrates to create bimetallic thin films using an apparatus with consistent parameters (such as temperature, solution flow rate, nozzle to substrate distance, and carrier gas pressure). Spray pyrolysis is used to prepare the Y-Fc, Mg-Al, and Mn-Zn oxide thin films. The metal precursor gels can be burned slowly at a high temperature to produce the bimetal oxide nano-powders. The smoldering process's leftover ashes are pulverized into an ultrafine powder [20].

3.3. Hydrothermal Method

High purity bimetallic oxides are produced by the hydrothermal synthesis. By placing the metal precursors in an autoclave under moderate heat and high pressure, this technique subjects them to forced hydrolysis. With the help of this technology, bimetal oxides with porous structural morphologies can be produced [22]. By adjusting the content of the metal precursors, the solvent's pH, the duration of the reaction, and the temperature of the autoclave, the structure of the bimetal oxide can be changed [23]. Bi-Nb, Cu-Co, and Sb-Mo oxides are synthesized and utilized in a variety of applications using the hydrothermal approach [24].

3.4. Sol-Gel Method

The sol-gel technique entails transforming a solution into a gel and then depositing it on a surface [25]. By first hydrolyzing the metal precursors and then polymerizing them, which creates a sol-gel, it is possible to create the bimetallic oxides with either mesoporous or microporous morphology [26]. Due to several factors such calcination temperatures, differences in the sol-gel procedure, and the concentration of the templates, bimetallic oxides with different structures may be produced [27]. Using the sol-gel process, the bimetal oxides of Zn-Fe, Co-Cu, and La-Fe are created [28, 29].

3.5. Impregnation Method

The easiest technique to create bimetallic oxides is by the impregnation procedure. An aqueous solution of metal precursor is impregnated into a metal oxide support that has already been prepared, and then calcinated. In a "wet" impregnation method, excess metal precursor solution is used to create thin slurry, while in a "dry" impregnation method; only a little amount of metal precursor solution is needed to fill up the support's pore volume [30]. The procedure of impregnation can be used to create the bimetal oxides of Mg-Fe, V-Cu, and Fe-Ni [31, 32].

4. Application of Bimetals and Bimetal Oxides for Waste Water Treatment

4.1. Removal of Toxic Metals and Heavy Metals from Waste Water by Bimetallic Nanoparticles

Various processes and reactions such as Redox reactions, dissolution, precipitation, desorption, and adsorption can influence the transformation, mobility, and solubility of heavy metals and also determine the toxicity of heavy metals in the environment [33]. Bimetals usually display tremendous redox activity in terms of removing heavy metal ions through their adsorption capacity, as shown in a recent water hygiene study by [34]. Iron-based bimetallic nanoparticles are used as some common adsorbents to remove heavy metals in water bodies. Removal of cadmium, arsenic, chromium, and mercury from waste water using bimetallic iron (Fe) based nanoparticles are among successful application in recent research. For instance, Fe-Ni nanoparticles were used for the removal of As (V) from mining effluents. Again, a calcined Fe-Ni bimetal was synthesized using eucalyptus leaf extract as a reducing agent. The adsorption process follows a spontaneous endothermic pathway and is well suited to the Langmuir isotherm model. The kinetics of the procedure was observed to follow a pseudo-second-order rate model, with a removal efficiency of 82.65%. Electrostatic interactions have been found to play an important role in the arsenic removal process [35].

Chromium (VI) can also be removed from wastewater using Cu-Fe bimetallic nanoparticles which are prepared by liquid-phase chemical reduction with an optimal Cu loading of 3% (w/w %). A mechanistic interpretation of the removal process by XPS (X-ray photoelectron spectroscopy) analysis revealed that Cr (VI) is reduced to Cr (III) and is concentrated on the surface of the particle as $Cr(OH)_3$ as shown in Figure 5. The reaction kinetics follows a pseudo-quadratic model and removal efficiency is decreased with an increasing pH [36].

Another recently published work in the Journal of Hazardous Materials in 2022 revealed that a bimetallic combination of copper and nickel was effective at removing pollutants such as heavy metals and organic compounds from wastewater. The study found that the bimetal had higher removal efficiency than either metal alone; a study published in the Journal of Cleaner production in 2020 found that a bimetal composed of zinc and copper was effective at removing pollutants such as lead and cadmium from waste water. The study found that the bimetal had higher removal efficiency than either metal alone, and that the bimetal was effective even at low concentrations, another study published in the Journal of Environmental Management in 2020 found that a bimetal composed of zinc and cobalt was effective at removing pollutants such as lead and cadmium from waste water at low pH levels [37, 38, 39].

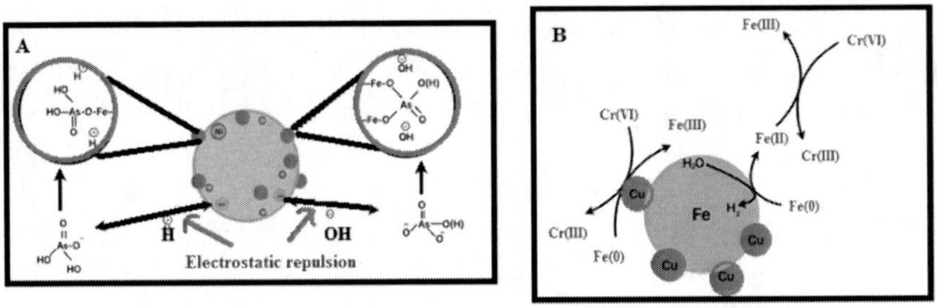

Figure 4. (A) represents Arsenic (V) removal mechanism, Figure B represent the Chromium (VI) removal from waste water by utilizing Cu-Fe bimetallic nanoparticle (Figures are adapted and recreated from (Allam, B.K., Musa, N., Debnath, A., Usman, U.L. and Banerjee, S., 2021. Recent developments and application of bimetallic based materials in water purification. *Environmental Challenges*, *5*, p.100405).

4.2. Removal of Heavy Metals by Bimetallic Oxide

Bimetallic oxides exhibit excellent adsorption capacity in removing heavy metal ions such as Co(II), Hg(II), and Cr(VI) from wastewater. The mechanism of remediation of metal ions from wastewater by bimetals differs from that of bimetallic oxides as they are commonly used as adsorbents in the process. A bimetallic oxide can reduce Hg from 11,600 ppb to 0.1 ppb in a single pass. The Zn-Sn-based metal oxides have been successfully used to remove Cu^{2+}, Pb^{2+}, and Sn^{2+} from contaminated water. The Langmuir model best fits the adsorption kinetics, with an adsorbent dose of 0.05 g and an initial concentration of 200 mg/l ion at 30 min to reach the maximum adsorption capacity of Cu^{2+}. The adsorption of Pb (II) ions on Fe-Mn oxide surfaces fits better with the polymolecular adsorption model. Adsorption mechanisms of Pb(II) on hydrated Fe(III) and Mn (IV) oxides have been described by Columbic, hydrolytic, ion-exchange, or coordinate sorption models and as a whole combination of all [40]. Other studies include the removal of heavy metal ions from aqueous solutions by bimetallic oxide nanoparticles" by Prabhakar et al. was reported in 2020. This study investigated the use of bimetallic oxide nanoparticles for the removal of heavy metal ions from aqueous solutions. The researchers found that the bimetallic oxide nanoparticles were highly effective at removing heavy metal ions, with a high removal efficiency for copper, lead, and zinc ions. Adsorption of heavy metals from wastewater using bimetallic oxide-coated magnetic nanoparticles was reported by Chmielewská, E. et al. This study investigated the use of bimetallic oxide-coated magnetic nanoparticles for the adsorption of heavy metals from wastewater. The researchers found that the bimetallic oxide-coated magnetic nanoparticles were highly effective at removing heavy metals such as lead, cadmium, and nickel from wastewater [40].

4.3. Use of Bimetals to Remove Anions and Oxyanion

The presence of a high level of anions and oxyanions can pose a serious threat to the environment and can have detrimental effects on life. Various uncontrolled human and industrial activities in the environment provide an access to these toxic anions and oxyanions and can pollute water resources. Anionic and oxygen-anionic trace contaminants commonly encountered in wastewater and surface waters are fluoride (F^-), nitrate (NO_3^-), phosphate (PO_4^{3-}), perchlorate (ClO_4^-), and sulfate (SO_4^{2-}). Mn (II)-La (III) bimetals can be synthesized by co-precipitation and are applied for strong pH dependent fluoride (F^-) decontamination. The sorbent exhibits a maximum adsorption capacity for fluoride (292.9 mg/g). The presence of coexisting anionic phosphate strongly inhibits the adsorption process. The Langmuir model was best suited for adsorption isotherms and implicit monolayer adsorption. The pseudo-second-order kinetic law best describes the experimental adsorption data with a correlation coefficient of r^2 =0.98 [41]. Other examples includes the successful utilization of Fe-Al bimetallic chitosan bentonite based (Fe-Al@bent) to remove nitrates and their derivatives from wastewater at low temperatures (2-7 degrees). The Fe-Al@bent removal efficiency of nitrate effluent at a concentration of 50 mg/L was approximately 90% in 60 min [42], application of rhenium-palladium bimetallic catalyst on activated carbon (Re-Pd@C) in the detection as well as reduction of perchlorate (ClO_4^-) to Chloride (Cl^-) under acidic conditions to regenerate brine from waste ion exchange (IX). Herein, the excess NO_3^- deactivates the catalytic activity of the ClO_4^- reduction process via interaction with the bimetallic Re (Rhenium) component. Further by pre-treating NO_3^- with In-Pd@Al_2O_3 the Rhenium deactivation can be avoided [43] etc. The electrochemical application of bimetallic systems in wastewater treatment is also reported. Examples like fabrication and application of Cu-Pt foam electrode to enhance the electrochemical reduction of nitrate in waste water with 94% conversion and 84% selectivity was reported in 2021. The investigation revealed that role of Cu surface supports the reduction of nitrate to nitrite and platinum nanoparticles facilitate the conversion of nitrite to ammonia [44]. Other studies include the application of needle-shaped Mg-La bimetallic oxide nanocomposites for the efficient removal of phosphate and fluoride with phosphate (107.34 mg/g) and fluoride (63.11 mg/g) and application of Fe-Ti bimetallic oxides as a best adsorbent for phosphate removal with 35.4mg/g at a pH 6.8 is recently reported [45].

4.4. Removal of Pesticides from Waste Water

Chlorinated organic compounds (COCs) are a class of hydrophobic organic pesticides used for the control of weeds and pests. These might get introduced into the aquatic ecosystem through processes like runoff from agriculture and industries, precipitation etc. Due to their lipophilic and recalcitrating nature, they are less prone to degradation and impose negative effects on both the environment and human health. Recent studies have proved that bimetal catalyzed dechlorination is a much efficient means of degradation of COCs present in wastewater [46]. The two methods used for COCs removal are summarized as follows:

4.4.1. Adsorption

The surface of bimetallic catalysts is generally heterogenous. So, the prevalence of adsorbate-adsorbent and adsorbate-adsorbate interactions must be considered during adsorption. Cooperative adsorption theory considers that the heterogenous catalyst surface has varied active sites with varied potentials thereby overcoming the limitations of BET and Langmuir Theory. The most common Fe-based bimetallic system is utilized for this purpose. Herein, a second metal M (such as Pd, Ni, Pt, or Cu) is deposited on the metal substrate and acts as a hydrodechlorination reagent. The active hydrogen species produced by the corrosion of the metal substrate is further stored in the M lattice. Moreover, the potential difference between the stable M cathode and the Fe anode can also boost reactivity as this induces more metal substrate corrosion [33]. Based on the dechlorination of dichlorobenzene over Pd-Fe bimetal, the explanation of cooperative adsorption theory [39] is included in this chapter. According to the theory, there are two types of adsorption sites on Palladium surface: "σ", the bigger active sites that can adsorb the aromatic compound and "σ*", smaller hydrogen adsorbing active sites. The complete reaction for hydrogen atom generation is shown below

$$Fe + 2H_2O + 2.\sigma^* \underset{K_H}{\overset{K_H}{\rightleftharpoons}} Fe^{2+} + 2OH^- + 2\sigma^*.H \quad (1)$$

Later the dechlorination mechanism occurs through two surface reactions shown in scheme A and Scheme B

$$OH^- + 2\sigma^*.H + \sigma.DCB \underset{K_{-1,C1}}{\overset{K_{1,C_1}}{\longleftrightarrow}} \sigma.CB + Cl^- + 2\sigma^* \quad (2)$$

Scheme A. Reaction between the adsorbed H atom and Dichlorobenzene (DCB).

$$OH^- + 2\sigma *.H + \sigma.CB \underset{K_{-1,C2}}{\overset{K_{1,C_2}}{\longleftrightarrow}} \sigma.B + Cl^- + 2\sigma^* \quad (3)$$

Scheme B. Reaction between Chlorobenzene (CB) and Benzene (B).

4.4.2. Reduction

Bimetallic nanoparticles mostly zerovalent iron removes various hazardous COCs such as carbon tetrachloride, trichloroethylene, tetrachloroethylene, nitroaromatics by reduction and also prevents the formation of toxic by-product. System like Fe-M, Fe is the reductant and M (Pd,Pt, Ag etc.) is the catalyst. Figure 5 visualizes the dechlorination mechanism on the bimetallic surface by reduction. Potential difference arises between Fe anode molecule and M cathode. Fe gets corroded to release H atoms which are adsorbed by M where hydrodechlorination (H-replacement) occurs by Cl-Cl bond cleavage [47].

However, due to the spontaneous and unregulated ZVI (Zero Valent Iron) corrosion, it is difficult to maintain the reactivity of Fe-based bimetals over a prolonged time period. Corrosion results in iron oxide precipitation, excess H_2 generation, and the loss of catalyst base, all of which contribute to the deactivation of M-Fe. Due to the possibility of a higher inherent reactivity and less mass transfer constraints brought on by smaller particle size, the reactivity of bimetallic NPs (nanoparticles)is significantly enhanced. However, because of their extremely high reactivity, metal oxide encapsulation or aggregation can quickly deactivate bimetallic NPs [33].

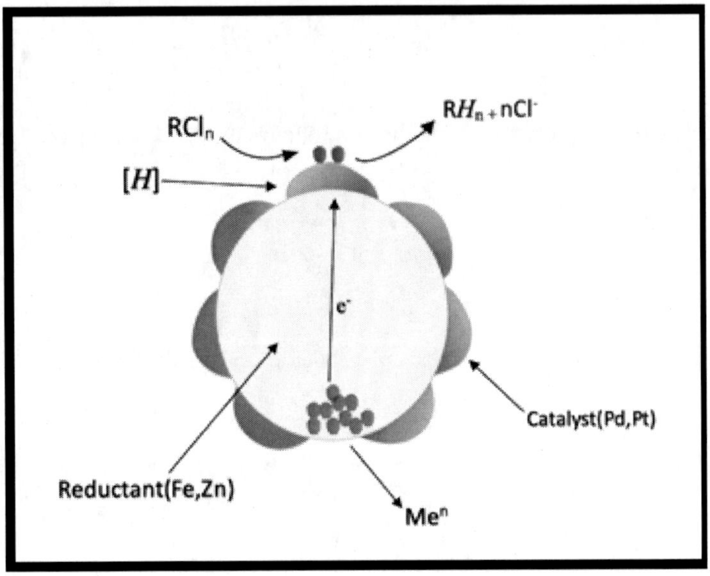

Figure 5. The pictorial representation of dechlorination on the heterogeneous surface of a bimetal. (Figure is adapted and recreated from the work of Zhang, W., Wang, C. and Lien, H., 1998; Catalytic reduction of chlorinated hydrocarbons by bimetallic particles. *Catal. Today*, *40*(4), pp.387-395).

4.5. Removal of Nitro Compounds and Azo Dyes

In general, nitro compounds are divided into nitro aliphatic and nitro aromatic compounds. Nitro aromatic chemicals are more extensively used and more difficult to breakdown than nitro aliphatic compounds, making them easier to detect in the environment. In particular, azo dyes, which have one or more azo (N=N) bonds, are the most significant and common family of synthetic dyes utilized in industries. Nitro or azo compounds are persistent pollutants because they include N-O and N=N linkages. Application of bimetallic oxide-based catalysts for the degradation of organic pollutants in water" recently reported by Liu and his group. This study proposed the development of bimetallic oxide-based catalysts for the degradation of organic pollutants in water. The catalysts were found to be highly effective at degrading pollutants such as methylene blue, bisphenol A, and methyl orange. "Synthesis and characterization of bimetallic oxide catalysts for treatment of dye wastewater" by Zhang and his coworkers. Their study reported the synthesis and characterization of bimetallic oxide catalysts for treatment of dye wastewater. The catalysts were found to be highly effective at degrading various dyes such as Methylene blue, Congo red, and crystal violet [48].

4.5.1. Use of Bimetal Catalysts for Azo Dyes and Nitrocompunds Removal

The use of bimetals was proved efficient in the degradation of stubborn nitro compounds. Fe-Cu and ZVI, in the presence of oxygen degraded p-nitrophenol into nontoxic and biodegradable intermediate, resulting in the release of CO_2 and H_2O as the final products [48]. For the removal of azo dyes, bimetals such as Fe-Ni, are successful catalysts for breaking the N=N .Reductive degradation of azo dyes such as Orange G, forming aniline and naphthol amine derivatives have been reported [49].

4.5.2. Use of Bimetal Oxide for Nitro and Azo Dyes Removal

Bimetal oxides are good degrading agents of azo dyes and nitro compounds as well. They work in a similar manner as the Bimetal catalysts. The azo link was the first to degrade, followed by the lower molecular weight intermediate being oxidized, and finally the complete mineralization of the dye into CO_2 and H_2O. In this framework, $CuO/\gamma\text{-}Al_2O_3$ catalyst showed remarkable reactivity. Various researches have been conducted for developing magnetically recoverable iron oxide bimetal catalysts. One such example is the utilization of Ag/Fe oxide composite nanoparticles for 4-nitrophenol removal. Novel palladium/hydroxyapatite/Fe_3O_4 nano catalyst exhibited high catalytic activity, magnetic separability and good stability for selective azo dye degradation [50]. The nitrophenols are being reduced catalytically to the matching amino derivative in a developing model reaction to access the catalytic activity. Herein, we have included the utilization of $NaBH_4$ and CuO/NiO bimetallic oxide nanoparticles as a catalyst to convert the stated nitro-phenols to their corresponding aminophenols which is one of the recently published works. Scanning electron morphology confirmed the production of uniformly distributed, spherical-sized bimetal oxide nanoparticles. The resulting bimetallic oxide nanoparticles showed remarkable crystal stability towards the nitrophenol reduction processes. The synthesized low-cost CuO/NiO NPs is an effective catalyst for converting harmful nitro compounds into beneficial by-products in both aqueous and non-aqueous media [51]. It's worth noting that these studies are published in peer-reviewed scientific journals and they are not conclusive evidence that the bimetals they study are the best solution for waste water treatment and the best way to apply them in real life scenario may vary based on the specific characteristics of the water, the pollutants, and other factors.

Conclusion

This chapter aimed to provide a summary of the synthetic methodologies and applications of bimetals and bimetallic oxides in waste water treatment. Recent development has created a great interest in the field of bimetal and bimetal oxide, their synthesis and application towards waste water treatment by removing the toxic pollutant. The catalytic properties of the bimetal and bimetal oxide varied accordingly to their synthesis procedures. Among physical and chemical method of preparation, chemical methods are often used. Methodologies associated with physical methods are complicated despite of environmental friendly. The most suitable method for bimetal oxides synthesis is precipitation method. This method is used to synthesize amorphous bimetallic oxides. One of the newly developed biosynthesis method utilizes the non-toxic reductants which are environmental friendly as well. Therefore, the synthetic approaches developed so far to prepare these compounds have been envisaged to expand for addressing the task-specific requirements. The range of applications of bimetal and bimetallic oxides discussed in this chapter has shown that the introduction of bimetal and bimetallic oxides ameliorates the green of the processes, leading to sustainable technologies, Although, the utilization of bimetal /bimetal oxides in elimination of toxic pollutant from water and waste water has gained boundless success, the mode of working is not same. In bimetal system, a substrate metal combines with a second metal that works as a potential alternative for reactions like hydrochlorination, hydrogenation as well as a storage tank for hydrogen. This makes them a suitable catalyst for the removal of toxic pesticides, azo dyes and nitro compounds from water. The high reducibility of bimetals provides the facility to remove metals and heavy metals like Cr(VI) as well as some anions. On the contrary, the bimetallic oxide is comprised of two metal oxides, one act as a host and the other oxide plays the functional key. The low cost and less reactive metal oxide play the role a host and the highly cost metal oxide plays the functional role with high reactivity. This facilitates them to work as versatile applicants like reductants/oxidant/catalyst in the elimination of toxins from waste water as well as other water systems. In conclusion, despite of numerous advantages, the major concern arises from the leaching of metal ions from the bimetal/ bimetal oxides which may be reason of high cost, reduction of catalyst life time as well as possible metal inferior to the preserved wastewater. So future study should definitely focus on the development of methodologies to decrease the amount of metal leaching as well as try to develop new straight forward and lost cost and less laborious synthesis methodologies as much as possible.

References

[1] Rodriguez, J. A. and Goodman, D. W., 1991. Surface science studies of the electronic and chemical properties of bimetallic systems. *The Journal of Physical Chemistry*, *95*(11), pp.4196-4206.

[2] Allam, B. K., Musa, N., Debnath, A., Usman, U. L. and Banerjee, S., 2021. Recent developments and application of bimetallic-based materials in water purification. *Environmental Challenges*, *5*, p.100405.

[3] Zaleska-Medynska, A., Marchelek, M., Diak, M., Grabowska, E., 2016. Noble metal-based bimetallic nanoparticles: the effect of the structure on the optical, catalytic and J. Scaria et al. 17 photocatalytic properties. *Adv. Colloid Interface Sci.* 229, 80–107. https://doi.org/ 10.1016/j.cis.2015.12.008.

[4] Wang, D., Li, Y., 2011. Bimetallic nanocrystals: liquid-phase synthesis and catalytic applications. *Adv. Mater.* 23, 1044–1060. https://doi.org/10.1002/ adma.201003695.

[5] Goia, D. V., Matijevic, E., 1998. Preparation of monodispersed metal particles. *New J. Chem.* 22, 1203–1215.

[6] Gilroy, K. D., Ruditskiy, A., Peng, H. C., Qin, D., Xia, Y., 2016. Bimetallic nanocrystals: syntheses, properties, and applications. *Chem. Rev.* 116, 10414–10472. https://doi. org/10.1021/acs.chemrev.6b00211.

[7] Gu, J., Zhang, Y. W., Tao, F., Feng, 2012. Shape control of bimetallic nanocatalysts through well-designed colloidal chemistry approaches. *Chem. Soc. Rev.* 41, 8050.

[8] Dabwan, A. H. A., Kaneco, S., Katsumata, H., Suzuki, T., Egusa, K., Ohta, K., 2010. Simultaneous removal of trihalomethanes by bimetallic Ag/Zn: kinetics study. *Front. Chem. Eng. China* 4, 322–327.

[9] Sun, Y., Xia, Y., 2004. Mechanistic study on the replacement reaction between silver nanostructures and chloroauric acid in aqueous medium. *J. Am. Chem. Soc.* 126, 3892–3901.

[10] Chau, J. L. H., Chen, C., Yang, C., 2017. Facile synthesis of bimetallic nanoparticles by femtosecond laser irradiation method. *Arab. J. Chem.* 10, S1395–S1401. https://doi. org/10.1016/j.arabjc.2013.04.014.

[11] Fu, F., Cheng, Z. and Lu, J., 2015. Synthesis and use of bimetals and bimetal oxides in contaminants removal from water: a review. *RSC advances*, *5*(104), pp.85395-85409.

[12] James, S. L., Adams, C. J., Bolm, C., Braga, D., Collier, P., Friscic, T., Grepioni, F., Harris, K. D. M., Hyett, G., Jones, W., Krebs, A., Mack, J., Maini, L., Orpen, A. G., Parkin, I. P., Shearouse, W. C., Steed, J. W., Waddell, D. C., 2012. Mechanochemistry: opportunities for new and cleaner synthesis. *Chem. Soc. Rev.* 41, 413–447.

[13] Tang, W. M., Zheng, Z. X., Tang, H. J., Ren, R., Wu, Y. C., 2007. Structural evolution and grain growth kinetics of the Fe–28Al elemental powder during mechanical alloying and annealing. *Intermetallics* 15, 1020–1026.

[14] Imawaka, K., Sugita, M., Takewaki, T., Tanaka, S., 2019. Mechanochemical synthesis of bimetallic CoZn-ZIFs with sodalite structure. *Polyhedron* 158, 290–295. https://doi. org/10.1016/j.poly.2018.11.018.

[15] Chen, J. C., Lin, Z. H., Ma, X. X., 2003. Evidence of the production of silver nanoparticles via pretreatment of Phoma sp.3.2883 with silver nitrate. *Lett. Appl. Microbiol.* 37, 105–108.

[16] Li, Y., Duan, X., Qian, Y., Yang, L., Liao, H., 1999. Nanocrystalline silver particles: synthesis, agglomeration, and sputtering induced by electron beam. *J. Colloid Interface Sci.* 209, 347–349.

[17] Thakkar, K. N., Mhatre, S. S., Parikh, R. Y., 2010. Biological synthesis of metallic nanoparticles. *Nanomed. Nanotechnol. Biol. Med.* 6, 257–262. https://doi.org/ 10.1016/j.nano.2009.07.002.

[18] Vilardi, G., Verdone, N., 2020. Production of metallic iron nanoparticles in a baffled stirred tank reactor: optimization via computational fluid dynamics simulation. *Particuology* 52, 83—96.

[19] Chai, L, Wang, Y., Zhao, N., Yang, W., You, X., 2013. Sulfate-doped Fe_3O_4/Al_2O_3 nanoparticles as a novel adsorbent for fluoride removal from drinking water. *Water Res.* 47 (12), 404T-4049.

[20] Chen, L., He, B. Y., He, S., Wang, T. J., Su, C. L., Jin, Y., 2012. Fe—Ti oxide nano-adsor- bent synthesized by co-precipitation for fluoride removal from drinking water and its adsorption mechanism. *Powder Technol.* 227, 3-8.

[21] Eslami, H., Ehrampoush, M. H., Esmaeili, A., Ebrahimi, A. A., Ghaneian, M. T., Falahzadeh, H., Salmani, M. H., 2019. Synthesis of mesoporous Fe-Mn bimetal oxide nanocomposite by aeration co-precipitation method: physicochemical, structural, and optical properties. *Mater. Chem. Phys.* 224, 65—72.

[22] Wangcheng, Z., Zhang, X., Yanglong, G., Li, W., Yuri, G., Guanzhong, L., 2014. Synthesis of mesoporous CeO_2-MnOx binary oxides and their catalytic performances for CO oxidation. *J. Rare Earths* 32 (2), 146—152.

[23] Ocsoy, I., Demirbas, A., McLamore, E. S., Altinsoy, B., Ildiz, N., Baldemir, A., 2017. Green synthesis with incorporated hydrothermal approaches for silver nanoparticles formation and enhanced antimicrobiai activity against bacterial and fungal pathogens. *J. Mol. Liq.* 238, 26S-269.

[24] Wang, J., Wang, B., Liu, Z., Fan, L., Zhang, Q., Ding, H., Wang, L, Yang, H., Yu, X., Lu, B., 2019. Nature of bimetallic oxide Sb_2MoO_6/rGO anode for high-performance potassium-ion batteries. *Adv. Sci.* 6 (17), 1900904.

[25] Vilardi, G., Stoller, M., Di Palma, L., Boodhoo, K., Verdone, N., 2019b. Metallic iron nanoparticles intensified production by spinning disk reactor: optimization and fluid dynamics modelling. *Chem. Eng. Process.-Process Intensif.* 146, 107683.

[26] Darmawan, A., Karlina, L., Astuti, Y., Wang, D., Motuzas, J., Da Costa, J., 2017. Interlayer free—nickel doped silica membranes for desalination. *IOP Conference Series: Materials Science and Engineering.* IOP Publishing.

[27] Polat, J. Vi., Soylu, A. M., Erdogan, D. A., Erguven, H., Vovk, E. I., Ozensoy, E., 2015. Influence of the sol—gel preparation method on the photocatalytic NO oxidation performance of TiO_2/Al_2O_3 binary oxides. *Catal. Today* 241, 2S—32.

[28] Feng, J., Kiu, T., Xu, Y., Zhao, J., He, Y., 2011. Effects of PVA content on the synthesis of LaFeO3 via sol-gel route. *Ceram. Int.* 37 (4), 1203-1207.

[29] Veith, M., Haas, M., Huch, V., 2005. Single source precursor approach for the sol—gel synthesis of nanocrystalline ZnFe2O4 and zinc— iron oxide composites. *Chem. Mater.* 17 (1), 95—101.

[30] Mehrabadi, B. A., Eskandari, S., Khan, U., White, R. D., Regalbuto, J. R., 2017. A review of preparation methods for supported metal catalysts. *Adv. Catal.* 61, 1—35.

[31] Tao, G., Hua, Z., Gao, Z., Zhu, Y., Chen, Y., Shu, Z., Zhang, L., Shi, J., 2013. KF-loaded mesoporous Mg—Fe bimetal oxides: high performance transesterification catalysts for biodiesel production. *Chem. Commun.* 49 (73), 8006—8008.

[32] Park, Y. M., Cho, J. M., Han, G. Y., Bae, J. W., 2021. Roles of highly ordered mesoporous smictures of Fe-Ni bimetal oxides for an enhanced high-temperature water-gas shift reaction activity. *Catal. Sci. Technol.* 11 (9), 3251—3260.

[33] Fu, F., Wang, Q., 2011. A review of removal of heavy metal ions from wastewaters. *J. Environ. Manage.* 92 (3), 407-418.

[34] Song, J., Huang, G., Han, D., Hou, Q., Gan, L., Zhang, M., 2021. A review of reactive media within permeable reactive barriers for the removal of heavy metal (loid) s in groundwater: current status and future prospects. *J. Clean. Prod.,* 128644.

[35] Lin, Y., Jin, X., Khan, N. I., Owens, G., Chen, Z., 2022. Bimetallic Fe/Ni nanoparticles derived from green synthesis for the removal of arsenic (V) in mine wastewater. *J. Environ. Manage.* 301, 113838.

[36] Ye, J., Wang, Y., Xu, Q., Wu, H., Tong, J., Shi, J., 2021. Removal of hexavalent chromium from wastewater by Cu/Fe bimetallic nanoparticles. *Sci. Rep.* 11 (1), 1-11.

[37] He, L., Yao, F., Zhong, Y., Tan, C., Hou, K., Pi, Z., Chen, S., Li, X. and Yang, Q., 2022. Achieving high-performance electrocatalytic reduction of nitrate by N-rich carbon-encapsulated Ni-Cu bimetallic nanoparticles supported nickel foam electrode. *Journal of Hazardous Materials, 436*, p.129253.

[38] Giannetti, B. F., Agostinho, F., Eras, J. C., Yang, Z. and Almeida, C. M. V. B., 2020. Cleaner production for achieving the sustainable development goals. *Journal of Cleaner Production, 271*, p.122127.

[39] Scaria, J., Nidheesh, P. V. and Kumar, M. S., 2020. Synthesis and applications of various bimetallic nanomaterials in water and wastewater treatment. *Journal of environmental management*, *259*, p.110011.

[40] Chmielewská, E., Tylus, W., Bujdoš, M., 2021. Study of mono-and bimetallic Fe and Mn oxide-supported clinoptilolite for improved Pb (II) removal. *Molecules* 26 (14), 4143.

[41] Yu, Y., Chen, J. P., 2014. Fabrication and performance of a Mn-La metal composite for remarkable decontamination of fluoride. *J. Mater. Chem.* A 2 (21), 8086-8093.

[42] Cheng, H., Zhu, Q., Wang, A., Weng, M., Xing, Z., 2020. Composite of chitosan and bentonite cladding Fe-Al bimetal: effective removal of nitrate and by-products from wastewater. *Environ. Res.* 184, 109336.

[43] Liu, X., Chen, Z., Chen, Z., Megharaj, M., Naidu, R., 2013b. Remediation of Direct Black G in wastewater using kaolin-supported bimetallic Fe/Ni nanoparticles. *Chem. Eng. J.* 223, 764-771.

[44] Cerrón Calle, G. A., Fajardo, A. S., Sánchez-Sánchez, C. M., Garcia-Segura, S., 2021. Highly reactive Cu-Pt bimetallic 3D-electrocatalyst for selective nitrate reduction to ammonia. *Appl. Catal. B,* 120844.

[45] Lu, J., Liu, D., Hao, J., Zhang, G., Lu, B., 2015. Phosphate removal from aqueous solutions by a nano-structured Fe-Ti bimetal oxide sorbent. *Chem. Eng. Res. Des.* 93, 652-661.

[46] Kong, L., Tian, Y., Pang, Z., Huang, X., Li, M., Li, N., Zhang, J., Zuo, W., Li, J., 2020.Needle-like Mg-La bimetal oxide nanocomposites derived from periclase and lan- thanum for cost-effective phosphate and fluoride removal: characterization, performance, and mechanism. *Chem. Eng. J.* 382, 122963.

[47] Zhang, W., Wang, C. and Lien, H., 1998. Catalytic reduction of chlorinated hydrocarbons by bimetallic particles. *Catal. Today*, *40*(4), pp.387-395.

[48] Xiong, Z., Lai, B., Yang, P., Zhou, Y., Wang, J. and Fang, S., 2015. Comparative study on the reactivity of Fe/Cu bimetallic particles and zero valent iron (ZVI) under different conditions of N_2, air or without aeration. *Journal of Hazardous Materials*, *297*, pp.261-268.

[49] Bokare, V., Jung, J. L., Chang, Y. Y. and Chang, Y. S., 2013. Reductive dechlorination of octachlorodibenzo-p-dioxin by nanosized zero-valent zinc: modeling of rate kinetics and congener profile. *Journal of Hazardous Materials*, *250*, pp.397-402.

[50] Chiou, J. R., Lai, B. H., Hsu, K. C. and Chen, D. H., 2013. One-pot green synthesis of silver/iron oxide composite nanoparticles for 4-nitrophenol reduction. *Journal of hazardous materials*, *248*, pp.394-400.

[51] Ramu, A. G., Kumari, M. A., Elshikh, M. S., Alkhamis, H. H., Alrefaei, A. F. and Choi, D., 2021. A facile and green synthesis of CuO/NiO nanoparticles and their removal activity of toxic nitro compounds in aqueous medium. *Chemosphere*, *271*, p.129475.

Index

T

U

X

Z